A SHORT HISTORY OF

Science and Scientific Thought

F. SHERWOOD TAYLOR was born in 1897 and educated at Lincoln College, Oxford and University College, London. He was a chemistry master at various secondary schools in England between 1921 and 1933, and then became an assistant lecturer in organic chemistry at Queen Mary College of the University of London.

Professor Taylor was Curator of the Museum of the History of Science at Oxford from 1940 to 1950, whereupon he took over the Directorship of the Science Museum in South Kensington, a post he held until his death.

Author of

THE WORLD OF SCIENCE
A SHORT HISTORY OF SCIENCE
A CENTURY OF SCIENCE
INORGANIC AND THEORETICAL CHEMISTRY
ORGANIC CHEMISTRY

A SHORT HISTORY

OF

Science and
Scientific Thought

With Readings from the
Great Scientists
from the
Babylonians to Einstein

By F. SHERWOOD TAYLOR, Ph.D.
Curator of the Museum of History of Science at Oxford

The Norton Library
W · W · NORTON & COMPANY · INC ·
NEW YORK

ACKNOWLEDGEMENTS

I wish to make grateful acknowledgement of the assistance of Professor B. G. Gunn in obtaining the photograph which appears as Pl. I, and to Dr. Hugh M. Raven for the photographs reproduced as Pl. XXII. My thanks are also due to the editors of *The Lancet* and the *British Medical Journal;* to the Liverpool University Press, to the Clarendon Press, to the University of Chicago Press, to the Princeton University Press, to Messrs. J. M. Dent and Sons, and to Messrs. T. Werner Laurie, for permission to reproduce extracts from their publications, the titles and authors of which are acknowledged at the foot of the extracts in question.

F. SHERWOOD TAYLOR

TIME CHART OF
MAIN PERIODS OF SCIENCE

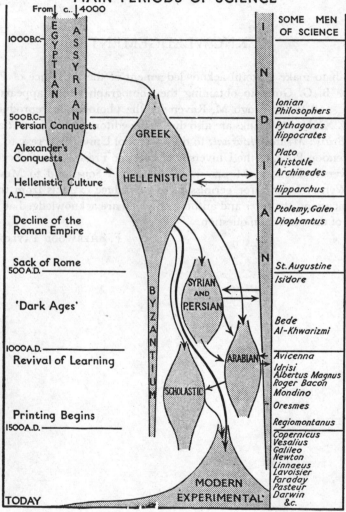

From c. 4000

SOME MEN
OF SCIENCE

1000 B.C.—

EGYPTIAN

ASSYRIAN

I N D I A N

Ionian
Philosophers

500 B.C.—
Persian Conquests

GREEK

Pythagoras
Hippocrates

Alexander's
Conquests

HELLENISTIC

Plato
Aristotle
Archimedes

Hellenistic Culture

Hipparchus

A.D.—

Ptolemy, Galen
Diophantus

Decline of the
Roman Empire

Sack of Rome
500 A.D. —

BYZANTIUM

St. Augustine

Isidore

SYRIAN
AND
PERSIAN

'Dark Ages'

Bede
Al-Khwarizmi

1000 A.D.—
Revival of Learning

ARABIAN

Avicenna
Idrisi
Albertus Magnus
Roger Bacon
Mondino

SCHOLASTIC

Oresmes

Printing Begins
1500 A.D.—

Regiomontanus

Copernicus
Vesalius
Galileo
Newton
Linnaeus
Lavoisier
Faraday
Pasteur
Darwin
&c.

MODERN
EXPERIMENTAL'

TODAY

CONTENTS

LIST OF PLATES

LIST OF FIGURES

A SHORT HISTORY OF

Science and Scientific Thought

CHAPTER ONE

The Study of the History of Science

The Importance of Science

It is not, perhaps, too much to say that Natural Science is the principal cause of human society being such as it is to-day. The very existence of perhaps two-thirds of the population of the world has been made possible by the application of science to the growing of food and the elimination of disease. The expectation of life has been multiplied by a factor of perhaps seven or eight in the last four centuries and this alone has altered the orientation of human thought. We do not expect to die young, and so we think more of how we shall live than how we may die. The external circumstances of our daily lives are almost wholly dependent upon science and there is scarcely anything in the modern home that has not been made by processes whose principles involve the scientific advances of recent years. Communications and transport have become so rapid as to bind the nations into an external intimacy in which they are very ill at ease. Power has been placed in man's hands by which he can work his will with unprecedented speed, and science has done nothing to prepare him ethically for its use. It is indeed transparently obvious that science has transformed the external world which man inhabits and uses: but it is also true that it has less directly transformed man's way of thinking. His approach to the age-old problems of his origin, his nature, his conduct and his destiny tend to-day to be scientific, which means that he looks for guidance more and more to that which can be stated as matters of scientifically observed fact, and less and less to the ethical side of his nature, which cannot be stated in such terms, but requires the use of the verbs 'ought' or 'should', which are unknown to science. The principal source of man's ethical and moral life has, throughout the ages, been religion, and, as a matter of experience, religion has declined where science has flourished—not necessarily, but from the failure of men to think clearly concerning the nature of religion and of science.

It is clear, then, that science has profoundly modified the lives of men; and this fact and its circumstances are quite obviously among the most important subjects that must be studied by the man who seeks to understand the world in which he finds himself.

Every citizen is directly and vitally concerned with science; for whether he know it or not, pure science has moulded his philosophy of life, and applied science has determined almost the whole of his material surroundings, his chances of life or death, poverty or affluence, freedom or slavery. Natural science is by far the greatest movement in modern history, and the most important factor to be taken into account in political planning.

The Nature of Science

In the preceding paragraph science is not defined, and for the historian it is not easy of definition, because through the ages it has been growing into an ever more sharply-defined entity. If we require a definition of science that shall be applicable from the earliest period of civilisation to the present time, we may say that *Science, in its widest sense, is a systematic method of describing and controlling the material world.* This is an activity which we find chronicled in those writings of the Egyptians and Babylonians that we call 'scientific', and it is the principal function of science to-day. Yet the activity to be described was necessarily prior to the description, and the roots of the activity that we call 'science' are in man's first dealings with the material world around him. A man who has acquired a craft has in his head and hands the laws of the behaviour of the material he works in, albeit unexpressed in words or writing: he has to subdue himself to the conditions that the material imposes and so he comes to establish a *true* relationship with it, a relationship which will be verified in the production of a satisfactory article. When certain crafts came to be studied by the learned and their methods recorded in writing, then was the beginning of recorded science. The progress from this beginning to the highly organised science of to-day is the subject of this book. We shall see how the practical recipes and records of the Egyptians and Babylonians gave place to the theoretical and philosophical science of the Greeks. We shall see how this, after enlightening and dominating the thought of the world for two thousand years, was rejected by the men of the seventeenth century, who divorced philosophy from science, which they refounded on a mathematical and quantitative basis, which has persisted essentially to the present day.

The Historical Study of Science

Science can be understood in two ways. It can be seen as it stands at any moment, as a logical coherent account of that order which the scientist of the time finds in nature; and the man who wishes to use science or to add to it can, in fact, get on very well without knowing its history, as we can see demonstrated in the persons of many scientists who are eminent in their subject, but little acquainted with its past.

But science is not merely a coherent system, here and now; it is also something that has grown and is growing, and that, as it grows, progressively affects man's life. Anything that grows and develops can only be understood by studying it historically. An organism, a movement, a civilisation must be studied in time; and in so far as we concern ourselves with science as something in progress, and as something that brings about changes in the institutions of mankind, so we must study its history. The responsible citizen of the world must come to an understanding of the way in which science influences and will influence philosophy, religion, and the externals of living: this he can accomplish only by noting the directions in which science has been modifying these matters in the past, and following those tendencies into the future.

The purpose of this book is then, to present the picture of science as an actively growing organism, and to that end we must return to the remotest beginnings of man's relation with nature.

CHAPTER TWO

The Beginnings of Science

Science and Crafts

The history of science, as we have seen, must begin with the history of crafts, for these are the foundation and necessary fore-runners of true science. Before architects and engineers come carpenters and smiths, and before these the simple un-specialised Man.

The beginnings of Man

The making of flint tools is the first known craft, which was discovered by creatures of a different species from ourselves, a million or so years ago, more or less. These early man-like creatures made rough flint weapons and worked bone and horn, but we know almost nothing as to their mentality. Our own species, *homo sapiens*, appeared perhaps some fifty thousand years ago, and from the first showed evidence of high intelligence. The first men had almost everything to learn: we do not even know if they had words by which to name and so to classify what they saw and handled. They improved the making of flint weapons; invented and brought to great perfection pictorial art. No one has drawn a bison better than it was drawn in the Altamira caves at the very beginning of the human story. Such drawings imply close observation, the making or collection of pigments, and also the use of fire and of lamps to see by, for they are depicted on the walls of dark caves.

Neolithic Culture

As time went on men began to build huts and to make crude pottery. Such cultures have existed in many different ages and parts of the world, but that which developed into civilisation and gave rise to science was in Egypt, where, perhaps about 5000 B.C. or even earlier, there dwelt the most cultured men of that distant age. They had learnt to domesticate animals and plants, and so were able to provide a securer supply of food and settle down in small communities. Dogs, cattle, goats, sheep and pigs were domesticated at least as early as the neolithic period immediately preceding civilisation in the near East: the ass probably somewhat later and the horse later still. Agriculture dates from a period at least as early, for wheat grains of a type showing much advance over the wild forms have been found in tombs and settlements dating from c. 3500 B.C. These men rapidly increased in skill. Their flint-work was such as no one could imitate to-day: they made stone vessels by grinding them out with emery; and although they had not the potter's wheel, yet they made sound and shapely pottery ornamented in black and red and white. They had learnt to make cloth and mats and baskets, they built simple huts to dwell in and travelled in boats with oars and sails: they even carved spirited representations of hunting and crude statues of their gods.

They lacked however three very important things, metal, writing, and a national organisation; and from the discovery of these we may date civilisation.

The Beginning of Civilisation

The very early history of civilisation is still largely unknown, but at the present it appears that in the earliest times there were at least three contemporary civilisations. First that of the Nile Valley—the Egyptian; secondly, that of the Sumerians, who transmitted their culture to the Babylonians in Mesopotamia and thence to the Assyrians; thirdly, a civilisation in the Indus valley, of which as yet we know comparatively little. All three civilisations seem to have developed about 4000–3400 B.C. and to have had a similar level of culture.

Material culture and scientific study did not continue steadily to advance throughout the thirty centuries in which the Egyptian and Assyrian cultures continued to flourish. It seems, on the contrary, that the first centuries of these civilisations were the greatest: that their art and learning reached the highest point before 2400 B.C. and thereafter were transmitted with no more than minor alterations, not always for the better. The products of the Old Kingdom of Egypt, made in the years following 3000 B.C., are artistically and technically equal or superior to the later ones.

Science before 600 B.C.

What did these early civilisations contribute to science?

(1) The necessary means for discovering and recording scientific fact; e.g. tools; vessels; materials of all kinds, and especially the metals; writing, and writing-materials.

(2) The beginnings of medicine and surgery.

(3) The beginnings of astronomy, and a fairly satisfactory calendar.

(4) The beginnings of mathematics.

This may not sound a great achievement for three thousand years of civilisation, but we have to remember, first, that beginnings are very hard to make, and that each advance in thought makes the next advance easier; secondly, that these peoples very soon found a way of living which was reasonably satisfactory to themselves, or at least to the learned caste, and that there was therefore little incentive to discovery.

The Necessities of Life

In the first few centuries of their civilisation the Egyptians invented a great many of the things which we no longer think of as inventions. Greatest of these perhaps was the development of tools, and the smelting of copper and bronze, which latter metal is a mixture of copper and tin.

Copper came into use probably before 4000 B.C. Ores containing copper carbonate or silicate are supposed to have been mined in the Sinai peninsula, and these could be smelted simply by burning them in charcoal-fires. The rough ingots so obtained were forged into tools by the hammer, and these tools enabled men greatly to improve their surroundings. With copper saws, fed with sand or emery, they cut the hardest stones, which they pounded and rubbed into shape and polished with infinite labour. Early Egyptian carpenters used bronze axes, adzes,

(1)

(2)

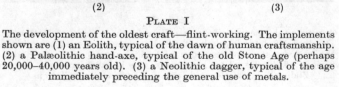

(3)

PLATE I

The development of the oldest craft—flint-working. The implements shown are (1) an Eolith, typical of the dawn of human craftsmanship. (2) a Palæolithic hand-axe, typical of the old Stone Age (perhaps 20,000–40,000 years old). (3) a Neolithic dagger, typical of the age immediately preceding the general use of metals.

PLATE II

A map of the heavens from a ceiling in the temple of Hathor at
Dendera. It dates from about the 1st century A.D., and combines
the 36 decan-stars (outermost circle) which mark out the 36 10-day
periods of the Egyptian year, with the Babylonian signs of the
Zodiac (inner circle) and certain Egyptian constellations.

saws, chisels, drills and knives: but arrowheads were still made of flint, for metal was too scarce to throw away. Commerce required balances and standard weights, as well as measures of length and means of reckoning. The use of metals originated what may have been the first specialised trade—that of the smith. The tools he made for the carpenter, the stonemason, and other makers of household goods, helped to elevate their activities into skilled trades; and the appearance of skilled tradesmen led in turn to more complex productions. The organisation of villages into bigger units created a wealthy class, which wanted better household furniture and equipment. So were made the first chairs and tables, and the same demand may have brought about the invention of the potter's wheel. Nobles and their ladies required beautiful jewellery; so the desert sands were washed for gold, and gold-workers vied with each other in the perfection of their wares. The rulers required palaces, nobler temples and more splendid and permanent tombs: these called for architecture and the beginnings of geometry, mechanics and surveying.

Supreme among these monuments is the Great Pyramid, built

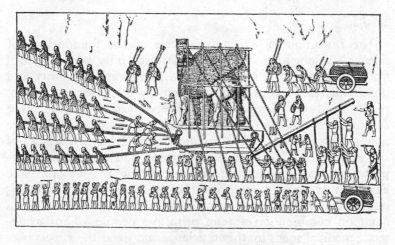

Fig 1.—Transporting a gigantic stone figure. Note use of rollers, levers and ropes—and unlimited man-power. (Neuburger, *Technical Arts and Sciences of the Ancients*. By courtesy of Messrs. Methuen.)

about 2800 B.C. as a king's tomb, and perhaps as something more. Its workmanship is incredible. Its sides, 254 yards long, differ only by two-thirds of an inch, and its angles differ from right angles by only a three-hundredth of a degree. Some of its stones weigh 50 tons, yet their edges are straight to within a hundredth of an inch and fit each other so that but a fiftieth of an inch of mortar separates them. As far as we know they were built simply by the concentration on this single task of enormous quantities of labour, skilled and unskilled. No more elaborate means of erection than wedges, ropes, levers, rollers and pulleys seem to have been employed. There are no pictures of the building of the Pyramids, but representations of the moving of colossal statues give us hints of how it was done. It is obvious that the Great Pyramid was not the work only of skilled labourers, but that architects, astronomers and mathematicians must have designed it and supervised the work. No written documents on these or other sciences have survived from that ancient time, but we have reason to suppose that some of the documents of a thousand years later are copies or adaptations of what was written in the great time of Egyptian discovery and achievement.

Writing

The invention of writing was one of the foundations of science. As long as knowledge was handed on by word of mouth from old to young, it had to remain comparatively scanty and unreliable, marvellous as is the memory of those who cannot read. Both the Egyptian and the Assyrian writings were complicated systems employing many characters, and before about 1000 B.C. they were the monopoly of learned scribes of the priestly class; wherefore all learning and much else of importance remained in the control of the priesthood. The Egyptian or Babylonian priest was not simply a 'clergyman'; but he was the man of learning and the man of God in one—a common combination among primitive people. Thus architects, medical men, mathematicians, and even the supervisors of metal-workers, were priests, and the temples were not only 'churches' but libraries, observatories and workshops.

The use of new materials

There was nothing that could strictly be called chemistry in Egypt and Babylonia, but there were crafts which had to do with materials, a great number of which were described and named. The metallurgists used and named many kinds of ores, and the common metals soon became available. Gold, being found as the native metal, was discovered before the beginning of civilisation. Copper followed; silver and bronze came into use a little later. Tin and lead were also known in early times; brass, however, does not seem to have been known until the Roman period. Iron comes into use from about 1500 B.C. but is not a common metal until about 850 B.C. The Sumerians and Babylonians were fine metal-workers. They also made coloured glasses and enamels, but transparent glass does not occur before about 1500 B.C. The making of glass implies a knowledge of alkalis and of the various metallic compounds which can be used as colouring material. The beautiful turquoise-blue Egyptian glaze is witness of a high technique, for such delicate temperature-regulation is needed for its firing, that only in recent years has it become possible to reproduce it. The Egyptians and Babylonians, in addition to these materials, record some hundreds of animal, vegetable and mineral products, which were employed as drugs. None the less they cannot be said to have begun the science of chemistry, for we have no record of any conjectures as to the manner in which these materials were related to each other.

Egyptian and Babylonian Medicine

The demand of the sick to be healed has always far outstripped the powers of the healer. Throughout history two methods have been employed, healing by mental means: magic, spells, dream-interpretation, etc.; and healing by physical means: drugs, baths, diet, manipulation, and the knife. Frequently both were employed together, but it would seem that the Babylonians and Assyrians excelled in spells, and the Egyptians in physical medicine. It must not be thought that mental healing is ineffective, for every doctor knows the power of faith. It is only when physical medicine and surgery become really efficient that mental healing is neglected; in ancient times many of the drugs and procedures were, we may suppose, useless and possibly harmful; and so the patient treated by spells was often better off in the hands of nature than he who was treated by drugs or manipulation.

Some very sound surgery was practised in Egypt, perhaps as early as 3000–2500 B.C. The papyrus named after Edwin Smith, of unknown authorship, written about 1500 B.C. but probably derived from sources at least a thousand years older, gives precise and well-founded instructions for treating such fractures, wounds and dislocations as could be treated successfully. In Babylon an unsuccessful or negligent surgeon might lose his hands, or at least be heavily fined, and the same may have been true in Egypt. In any event, the papyrus carefully distinguishes ailments which may be treated, from those that are likely to be fatal and should be left alone (p. 17). The Edwin Smith papyrus is extremely practical and unlike most of the documents of this time is quite free from superstition.

The medical papyri do not impress us so powerfully, but we must remember that the assessment of the value of drugs is a very difficult task; and, considering that a number of totally useless drugs were still employed in the nineteenth century A.D., we must not expect too much of the sixteenth century B.C.

The Beginnings of Astronomy

Some knowledge of the heavenly lights is needed by even the most primitive people; for they must reckon time, and their only timekeepers are the sun, and moon, and stars, which measure out the day, month, and year. The beauty and glory of the sun and its effect on living nature, are such that almost all races, the Egyptians and Babylonians included, have identified it or associated it with a god: the moon and planets, likewise, are commonly linked with spiritual powers. Astronomy was therefore studied, both to provide a calendar, and to predict natural and human events from the movements of the celestial powers; this prediction we now call astrology, but in early times this cannot be distinguished from astronomy. The calendar was not only practical but religious, for yearly events such as seed-time and harvest are commonly associated with religious festivals. In Egypt, where the whole of agriculture turns upon the cardinal date of the rising of the Nile, the prediction of this event from the positions of the stars was obviously of the first importance.

We do not know how long it took to settle the outlines of the calendar, that is to count the days in a month and the days and months in a year. To do this it is necessary to mark the instants at which the day, month, or year, begins and ends. Days could be counted from noon to noon, this being the moment when the shadow of a vertical stick is shortest. Months could be counted, e.g. from the first appearance of the crescent moon till its next first appearance, but this could only indicate the beginning and end of the month to the nearest whole day. The year could be reckoned from solstice to solstice. In winter the sun rises further to the South of East each day, until mid-winter, when its place of rising begins to return towards the North. The day on which it reaches the most southerly point is the winter solstice, and a year is the period from one solstice to the next. The year could also be reckoned from equinox to equinox, this being the day when the rising sun, the setting sun, and the point of observation lie in one straight line. By counting great numbers of days, months and years, and by dividing the number of days by the number of months or years, the Egyptians and Babylonians settled the year first as 360, then 365, and finally as $365\frac{1}{4}$ days. The month was taken as $29\frac{1}{2}$ days, so they had months of alternately 29 and 30 days, and the astronomers thus contrived that full moon always came at the middle of the month. Both the month and the year were correct to a few minutes, but as 12 lunar months are only 354 days, every two years or so an extra month was inserted into the year in order to keep the same month at about the same time of the year. The astronomers who regulated this and also made astrological predictions for the King are "the astrologers, the stargazers, the monthly prognosticators" of Isaiah 47, 13.

The Egyptians deliberately made no allowance for the quarter of a day over and above the 365 whole days which made up this year, so every Egyptian year started a quarter of a day

too soon, and their calendar was like a clock that gains. All
the fixed feasts gained on their true times and ran through all
the seasons in turn: and just as a clock that gains comes right
again in time, so after 1461 years, which they called a Sothis,
a feast which had been at, let us say, the spring equinox came
back to the spring equinox again. It was an Egyptian astronomer, Sosigenes, that advised Julius Cæsar in constructing
the Julian calendar in which the extra quarters of a day in four
successive years were gathered up and made an extra day of
leap-year, the last of the four.

The Egyptians or Babylonians divided the sky into the twelve
signs of the Zodiac, named most of the constellations, and discovered all the five planets which were known before the
eighteenth century. Astronomical records in Babylonia go back
to a very remote era. There is evidence of the recording of the
eclipse of March 8th 2283 B.C. which was believed to presage
the fall of the city of Ur, and the beginnings of astronomy in
Mesopotamia can be dated as near 4000 B.C. From the reign
of Nabonassor (747 B.C.) up to about A.D. 100, the 'Chaldæan'
astronomers kept continuous records of observation. It seems
that the older Babylonian astronomy was mainly astrological.
It was very important to them to be able to predict eclipses.
They recorded eclipses of the moon and sun for religious reasons,
and discovered that they recurred at intervals of a Saros, about
18 years (223 lunations), so that they were able to predict them
moderately well. These eighteen-year cycles give all the times
when the sun, moon and earth are in line, and it is only at these
times that an eclipse is possible: it often fails to take place at
these times because their orbits are not in the same plane. But
the astronomers did not mind predicting an eclipse that did
not in fact take place, for this meant that the gods had averted
the portent: they were, however, subject to censure and peril
if an eclipse occurred which they had failed to predict. From
about 500 B.C. these astronomers acquired more accurate
methods than the use of the Saros; in fact their later predictions could not have been made by the use of any cycle, however accurate. The later Chaldæans must have been able to
compute positions from the elements of motion, as do modern

astronomers, and so accurately that their figures for the motions of the Sun and Moon have only about three times the error of the best astronomers of the eighteen-fifties. We shall have more to say of this work apropos of Greek astronomy.

At night, time could be measured by the stars, which move through one sign of the Zodiac in two hours: in the day, time was reckoned by the sundial. The Babylonians probably invented this—in the form of a vertical stick casting a shadow: they also measured time by water-clocks, vessels from which water slowly leaked away and in which the water-level marked the hour. These are inaccurate because the viscosity and surface-tension of the water, which determine the rate of flow, vary with its temperature; but they remained the only timekeepers independent of the heavens until c. A.D. 1250, when the first mechanical clocks seem to have been invented: sand-glasses did not come in until the fifteenth century. Indeed, no clock that was more accurate than a sundial was available until the second half of the seventeenth century.

The beginning of Mathematics

We may suppose that counting began long before civilisation, but we have no reason to suppose that arithmetic existed before the Old Kingdom of Egypt. The Egyptians and Babylonians were a commercial people. They bought and sold goods, made buildings, owned land and reckoned time, so they required an arithmetic capable of handling the problems of commerce and astronomy, and a geometry sufficient for the surveyor and architect. They needed, for example, to work out how to share goods in various proportions and to know how tall and wide to build a granary to hold a given number of measures of corn. This they learned to do, but, as far as we know, they were interested in mathematics as a practical art, but not as an intellectual pleasure or discipline; and they would never have built up a system of geometry for the sheer interest of it, as did the Greeks.

Counting usually begins with the fingers and toes. Most primitives count up to five or ten or twenty and then start again. The Egyptians counted by tens; their arithmetic was not as easy as ours because they had quite different signs for

10, 100 and 1000 (as in Roman figures), instead of using the
same sign and altering its position as we do. They could add,
subtract, multiply and divide, which latter they did by repeatedly doubling or halving. The Babylonians, who were the
finest mathematicians of those ancient times, were exceptional in counting both by tens and by sixties. This was a
learned system, artificially compounded of the primitive
counting by tens and the convenient number 6. Thus 60 is
divisible by 2, 3, 4, 5, 6, a great convenience in calculation.
The Babylonians employed three signs, the wedge ▼ that
signified $\frac{1}{60}$, 1, 60, 60², etc., according to position, and the
sign < which meant 10, and the zero, ≷ which could be used
internally (as in 105), but not externally (as in 150). Thus
▼▼▼≷>> ▼▼ can be written 3×60²+0×60+22 or 10822.

Civilisations other than in Egypt and Mesopotamia

As far as we know, the Egyptians and Babylonians were by
far the most advanced peoples of the years 4000–600 B.C. Very
little is known about India and China at this time: undoubtedly
there were independent civilisations in both these countries
before 1000 B.C., but much excavation and study will be required before we know what they achieved. In the years
between 2000 and 600 B.C. many other centres of civilisation
derived their culture partly or wholly from Egypt or Babylon.
The Cretans, Phœnicians, Jews, and Hittites may serve as
examples, but none of these seem to have made noteworthy
contributions to science. The next wave of culture appears
among the Greek-speaking people in Greece and its colonies,
Asia Minor and Southern Italy.

Examples of Egyptian and Babylonian Science

BABYLONIAN SOLUTION OF A QUADRATIC EQUATION, C. 1700 B.C.

(The problem is stated and worked in the scale of 60. So 1
is unity, as in the usual denary scale, but 15' is $\frac{15}{60}$, i.e. ¼,
and 30' is ½ and 45' is ¾.

The problem is that of finding the side of a square from the data given.)

I have added the surface and the side of my square: 45'.

You will put 1, unity. You will break one in two: 30'. You will multiply 30' and 30' : 15'.

You will add 15' to 45' : 1. That is the square of 1. You will subtract 30', which you multiplied, from 1 : 30', the side of the square.

(The problem can be written $x^2 + x = \frac{3}{4}$.

The scribe squares a half and adds it to both sides

$$x^2 + x + \tfrac{1}{4} = \tfrac{3}{4} + \tfrac{1}{4}$$
$$(x + \tfrac{1}{2})^2 = 1$$
$$x + \tfrac{1}{2} = 1$$
$$x = 1 - \tfrac{1}{2}$$
$$\dots \quad x = \tfrac{1}{2}$$

The Babylonians solved much more complicated quadratics than this, and indeed had a general solution of the equation of the type $ax^2 + bx = c$, very similar in principle to our own. It is to be noted that the Babylonians do not use algebraic notation nor do they write an equation.)

(Adapted from *Textes Mathematiques Babyloniens*.

F. Thureau-Dangin, Leiden, R. J. Brill, 1938.)

EGYPTIAN MATHEMATICS (C. 1700 B.C.)

Example of reckoning the produce of a herdsman. Behold now this herdsman came to the numbering of cattle with 70 oxen: Said the accountant of cattle to this herdsman, How few are the head of oxen which thou hast brought! Where then are thy numerous head of oxen? This herdsman said to him, What I have brought thee is two-thirds of one-third of the cattle which thou didst entrust to me. Count for me and thou wilt find me complete.

The doing as it occurs:

1	1	1	1/6 + 1/18	Multiply 70 by 4½
2/3	2/3	2	1/3 + 1/9	Result 315: these are what were en-
1/3	1/3	4	2/3 + 1/6+1/18	trusted to him.

2/3 of 1/3 of it is 1/6 + 1/18 ½   1/9   1   315
    Divide 1 by 1/6 + 1/18   Total   1   2/3   210
                                              1/3   105
                          2/3 of 1/3 it is 70:
                          these are what he brought

(The question is: *If two-thirds of one-third of a number is 70, find the number.* Egyptian arithmetic was chiefly a collection of facts about numbers, like our multiplication table. He had hardly any general rules, but one of the very few of these is used in this example. By a rule which we would express as $2/3$ of $1/x = 1/2x + 1/6x$: they could reckon that 2/3 of 1/3 was $1/6 + 1/18$. This is inverted; that is to say, 1 is divided by it, giving $4\frac{1}{2}$, which, multiplied by 70, is 315: the rest of the sum proves that answer to be correct. The calculation is taken from the *Rhind Mathematical Papyrus*, translated by T. Eric Peet, University Press of Liverpool, 1922, page 110. No. 67.)

EGYPTIAN SURGICAL INSTRUCTION CONCERNING BROKEN NOSES

Instructions concerning a break in the Chamber[1] of the Nose

If thou examinest a man having a break in the chamber of his nose, (and) thou findest his nose bent, while his face is disfigured[2], (and) the swelling which is over it is protruding Thou should say concerning him: "One having a break in the chamber of his nose. An ailment which I will treat."

Thou shouldst force it to fall in, so that it is lying in its place, (and) clean out for him the interior of both his nostrils with two swabs of linen until every worm of blood[3] which coagulates in the inside of his nostrils comes forth. Now afterwards thou shouldst place two plugs of linen saturated with grease and put into his two nostrils. Thou shouldst place for him two stiff rolls of linen, bound on. Thou shouldst treat him afterwards with grease, honey, (and) lint every day until he recovers.

Instruction concerning a smash in the nostril

If thou examinest a man having a smash in his nostril, thou

shouldst place thy hand upon his nose at the point of the smash. Should it crepitate under thy fingers, while at the same time he discharges blood from his nostril (and) from his ear, on the side of him having the smash; it is painful when he opens his mouth because of it, (and) he is speechless.

Thou should say concerning him: "One having a smash in his nostril. An ailment not to be treated."

Notes by a commentator about 2600 B.C. to whom this treatise was even then ancient enough to require explanation.

1. "A break in the chamber of the nose", it means the middle of his nose as far as the back, extending to the region between the two eyebrows.

2. As for: "His nose bent while his face is disfigured", it means that his nose is crooked and greatly swollen throughout; his cheeks likewise, so that his face is disfigured by it, not being in its customary form, because all the depressions are clothed with swellings, so that his face looks disfigured by it.

3. As for "every worm of blood which coagulates in the inside of his two nostrils", it means the clotting of blood in the inside of his two nostrils, likened to the '*n*'.*r.t.*—worm, which subsists in the water.

(*The Edwin Smith Surgical Papyrus*, T. H. Breasted, Chicago, 1930. Vol. I.)

REPORT OF BABYLONIAN ASTRONOMER-ASTROLOGERS

No. 162 (Rev. 2ff.). When Jupiter goes with Venus, the prayer of the land will reach the heart of the gods. Merodach and Sarpanitum will hear the prayer of thy people and will have mercy on thy people.

Let them send me an ass that it may ease my feet. From Nirgal-itir.

REPORT OF A PREDICTED ECLIPSE WHICH THE OBSERVER THINKS MAY NOT HAVE IN FACT OCCURRED

No. 274. To the king of countries, my lord thy servant (Bilusur (?) May Bel, Nebo and Samas be gracious to the king my lord. An eclipse passed the city of Assur, wherein the king is dwelling; now there are clouds everywhere so that whether it did or did not happen we do not know. Let the lord of kings send to Assur, to all cities, to Babylon, Nippur, Erech and Borsippa; whatever has been seen in those cities the king will hear for certain. The omens (?) . . . the omen for an eclipse happened in Adar and Nisan; I send all to the kings, my lord, and they shall make a nambulbi-ceremony for the eclipse. Without fail (?) let not the king omit (?) to act rightly. The great gods in the city wherein the king dwells have obscured the heavens and will not allow the eclipse; so let the king know that this eclipse is not directed against the king, my lord, or his country. Let the king rejoice. . . .

(*The Reports of the Magicians and Astrologers of Nineveh and Babylon*, by R. C. Thompson, Luzac, 1900. Vol. II.)

BABYLONIAN EPHEMERIS

This tablet, for the month of Nisan (about Easter) 123 B.C., gives the position of the moon and also the following information about the planets.

Nisannu   8 Mercury goes to the West into Taurus heliacally. On night 10 in the evening Mars under β-geminorum 2 'yards'.* 10 Jupiter goes into the beginning of Taurus heliacally. During the night 13 in the evening Mercury under α-tauri 4 yards. 14 Saturn goes into the beginning of Aries heliacally. On night 21 in the evening Mercury under β-tauri 1 yard 6 inches. Night 23 in the evening Mercury under ζ-tauri 1 2/3 yards. Night 30 in the evening Mars under γ-canceri 4 inches.

(From F. X. Kügel S. J. *Sternkünde und Sterndienst in Babel*.)

---

*The word translated yard signifies an angle of 2°5', that translated "inch", 6'25"

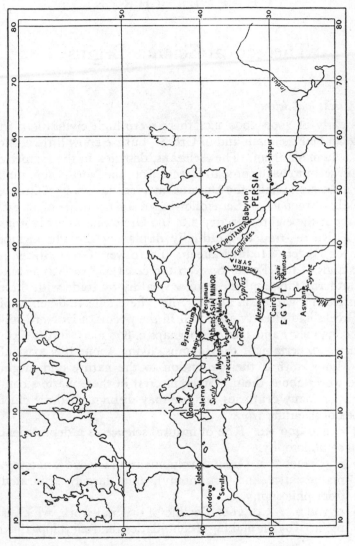

FIG. 2.—Map showing some centres of scientific activity before A.D. 1000.

CHAPTER THREE

Theoretical Science Begins

The Greeks and Science

As early as 3000–2000 B.C. there were high civilisations in Crete and on the mainland of Greece, but we know little of the men who made them. The Achæans, depicted in the Homeric poems, seem to have come to Greece from the North some time after 2000 B.C.: finally the Dorians invaded the whole of Greece from about 1200 B.C. There followed several centuries of unrest, but about the eighth century B.C. the Greeks became vigorous and active sea-traders, gradually displacing the Phœnicians from the Eastern Mediterranean. There were Greek States in the islands of the Ægean Sea, on the coast of Asia Minor and in Southern Italy, and these grew wealthy by trade with these countries and with Asia and Egypt. The intellectual genius of the Greeks had already appeared in the poems of Homer (perhaps c. 950 B.C.) and Hesiod (perhaps c. 800 B.C.): in the sixth century B.C. certain of the Ionians, living about the coast of Asia Minor, turned their attention to the nature and origin of the world about them. Greek interest in these matters continued for many centuries and we may distinguish three chief periods of scientific work.

(1) c. 600–400 B.C. Rise of natural science as a department of philosophy.

(2) c. 400–200 B.C. Great developments in science (anatomy, mathematics and astronomy), now largely separated from philosophy.

(3) c. 200 B.C.–A.D. 200, Decline of original work, which is replaced by the making of compilations and encyclopædias.

The spirit of enquiry

The contribution of the Greeks was nothing less than the creation of the very idea of science as we know it. As far as we know, the Egyptians and earlier Babylonians recorded and studied only those facts about the material world that were of immediate practical use, whereas the Greeks introduced what is still the chief motive of science, the desire to make a mental model of the whole working of the universe. Practical use they despised, and they desired the knowledge of things as a means of understanding and realising the harmony and order of the world. It is one thing to desire knowledge: another to find the right way to achieve it, and the first requisite for obtaining accurate knowledge about the world, namely the making of great numbers of accurately recorded observations and experiments, was not to the taste of the Greeks. They were full of curiosity: they had great artistic ability: but they always preferred the discussion of abstract principles to those practical measurements and weighings and prosaic descriptions of things that are the material of science. Simple experiments with tools and vessels and mechanical contrivances they felt to be slavish and degrading, so naturally they did not go far with physics and chemistry. Certain manual operations, it is true, had an honourable origin and tradition: the healing art, and to a less extent the study of animals and plants, was not unworthy of a philosopher, while astronomy, the study of the eternal mathematical harmony of the universe, was nobler still. But the greatest success of the Greeks was in geometry, for which little or no observation or experiment was required, but simply the exercise of pure reason.

The Greeks found few obstacles to scientific progress other than those inherent in their own temperament. There was little in the way of religious dogma with which their scientific speculations could come into contact; and, although Anaxagoras was brought to trial on the charge of asserting that the sun might be an inanimate mass of red-hot rock, it must be said that, considering the boldness and width of their discussions, the Greek philosophers were little troubled by religious restrictions. The Greeks were always great talkers and free speculators, critical and humorous: so it is not surprising that in the first two hundred years of their science they managed to discuss a great many fundamental scientific problems.

The Ionians and their critics

Modern science starts with small, readily soluble problems; It asks how things fall or float or melt, and comes lastly, if at all, to the great problems of the nature of life and the universe. Greek science started with the most difficult of questions— how the world came into existence. Before their time the answer had been a *creation-myth*, the story of how a personal God or Gods had fashioned the world, sometimes from nothing, sometimes from some simple formless matter such as water or slime. The Ionian philosophers likewise pictured the complex world they knew as having arisen from something simple and uniform, but they pictured it as having arisen *by natural causes*. They did not deny the existence of gods, but they tried to explain how the world might have come to be through impersonal agencies. The works of the three early Ionian philosophers, Thales, Anaximander, Anaximenes are lost; but we know that they all agreed that the world had come from one *simple stuff*. Thales called it '*water*', Anaximander '*the indefinite*', Anaximenes, '*air*': but we must be careful not to read any of our scientific ideas about air and water into their work. Next they thought that this mass of uniform "first principle" separated into parts, some hotter and some colder, some heavier and some lighter, and this difference of parts brought about circulation and motion, the result of which was the series of changes that have brought the earth to its present state. To-day, as far as we know anything of the origin of the universe, we suppose it to have arisen from condensation of a homogeneous cloud of cosmic dust, so the Ionians were not so far from an aspect of what we now think to be the truth.

At about the same period Pythagoras and his followers originated the idea of *the mathematical character of the universe*. They were preoccupied with numbers and they seem to have pictured the world as a structure of geometrically arranged points, just as we see it as geometrically ranged 'lattices' of atoms. Pythagoras, who was pre-eminently a mathematician, was not interested in the *kind* of stuff the universe was made of, but in its harmony and proportions. The Pythagorean believed that if he knew the mathematical relationships that lay behind the visible universe he would understand it.

These theories are static, they are concerned with what things *are* rather than with how they *move*. Heracleitus (fifth century B.C.) was the first to be impressed with the ever-changing character of the universe: it was ever flowing, a living fire, in which the only rest was that of balanced forces, as of the bow and the string. Parmenides on the other hand had great difficulty in seeing how change could come about at all and concluded that it was an illusion of our senses, the real world being externally unchanging.

Empedocles contributed the notion that all things were made of *elements*, root-substances of which all the varying objects of the world were compounded: this notion of elements is one that has come down to modern times.

So the earliest Greek Philosophers, of whom we know very little because most of their works are lost, contributed fundamental and permanent notions about the universe, but very little in the way of positive scientific fact, except in the fields of geometry and perhaps astronomy.

Plato and Aristotle

These two great men, who lived in the fourth century B.C., had the deepest influence on the world's thought, scientific and otherwise. The actual scientific facts they record are not very numerous and largely wrong, but they taught the world how to reason. As we shall see, the Greeks had invented geometry (as distinct from mere rules-of-thumb about measurement), and they were greatly impressed with the power of geometry to discover and prove all kinds of properties of complicated figures, such as spheres and cones and polyhedra, simply by starting with a few simple definitions and reasoning correctly about them. So they came to take geometrical reasoning as a model for scientific reasoning in general: this would have done very well, if they had had accurate observations to reason about, but, in fact, they did not realise the difficulty of ascertaining the truth about the simplest things.

Plato and Aristotle saw that scientific knowledge was not of individual things but of classes. The circle in a geometrical

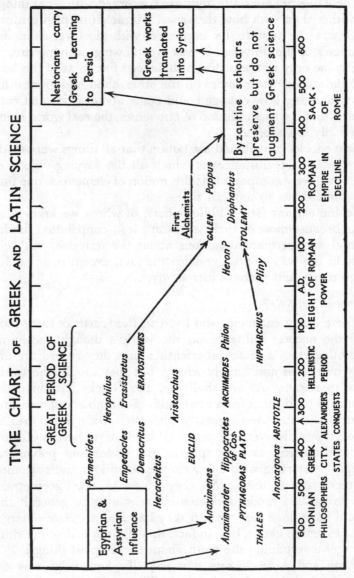

TIME CHART OF GREEK AND LATIN SCIENCE

Egyptian & Assyrian Influence

Nestorians carry Greek Learning to Persia

Greek works translated into Syriac

Byzantine scholars preserve but do not augment Greek science

First Alchemists

GREAT PERIOD OF GREEK SCIENCE

Parmenides
Herophilus
Empedocles   Erasistratus
Democritus   ERATOSTHENES
Heracleitus   Aristarchus
GALEN   Pappus
Diophantus
Heron ?   PTOLEMY
Pliny
ARCHIMEDES   Philon
Hippocrates   HIPPARCHUS
OF COS
Anaximander   PYTHAGORAS   PLATO
Anaximenes   EUCLID
Anaxagoras   ARISTOTLE
THALES

600   500   400   300   200   100   B.C.   A.D.   100   200   300   400   500   600

IONIAN   GREEK   ALEXANDER'S   HELLENISTIC   HEIGHT OF ROMAN   ROMAN   SACK.
PHILOSOPHERS   CITY   CONQUESTS   PERIOD   POWER   EMPIRE IN   OF
    STATES   DECLINE   ROME

FIG. 3.

proof was not any particular circle but the ideal circle. So the doctor studies the Body—not the body of Mr. A. or Mrs. B., but the body in general. Throughout history men have discussed the relationship between these universals and particulars. Plato would say that the circle we study in geometry was a real, unchangeable, eternal existence, and that the circles we draw on paper or visualise are imperfect copies of it. Aristotle would have the latter circles to be real, and the circle about which we make assertions in geometry to be a mental abstraction only. Because of such views Plato was not much interested in scientific observation of individual things, while Aristotle on the other hand made some admirable observations in biology.

The Elements

The Ionians thought the world sprang from a single substance, e.g. water or air, but Empedocles and many later writers thought there were and had been from the first several elements or 'roots of things' in matter. Plato adopts this view and believes his four elements, earth, water, air and fire, differ because their atoms have different shapes. Aristotle did not believe in atoms, because he did not think a void was possible, and so could not believe in the empty space which must be supposed to lie between atoms. He regarded everything as *matter + form*, form being the hidden cause of what we now call the properties of a thing, and matter the stuff that has these properties. Neither matter nor form could exist alone, but the simplest combinations of matter with the minimum of form were his four *elements*.

Earth.	dry and cold	
Water.	moist and cold	prime matter
Air.	moist and hot	
Fire.	dry and hot	

Unlike our elements these could change into one another, for they were all modifications of simple formless matter. They did not represent what we call earth, water, air and fire, but more nearly the qualities of

(1) the heavy, infusible and solid
(2) the fusible and liquid
(3) the gaseous and volatile
(4) the volatile, energetic, and light

But as there was no way of discovering with certainty which elements and what proportion of them any specified piece of matter contained, this theory was not of practical use to those who were concerned in the manufacture and use of different kinds of matter.

Greek Astronomy

The Greeks certainly took over a good deal of knowledge of the stars and planets from Egypt and Babylon. It is difficult to conjecture how much the earliest Greeks learnt from their predecessors, but it seems that the highly accurate knowledge of the Babylonians concerning the motions of the heavenly bodies was available to the Greek astronomers from the fourth century B.C. We must think of the Greeks and these Chaldæan astronomers as working simultaneously along rather different lines. The Chaldæans had made an enormous mass of observations over a very long period and so could discover the regularities in the solar, lunar and planetary motions with great exactness: they also had mathematics very suitable for numerical calculation. The Greeks, on the other hand, had fewer observations and those not very accurate; they were not great computers but were wonderful geometers, and so they excelled, not in making accurate calculations concerning the motions of the heavenly bodies, but in finding geometrical explanations of them.

The earliest Greeks seem to have thought the earth to be flat or cylindrical, but, from the fifth century on they knew that the earth was a sphere, and nearly all of them supposed this sphere to be in the centre of the universe and motionless. The stars were supposed to be at the surface of a vast sphere concentric with the earth. The stars rise and set once a day, so this sphere was supposed to rotate on its axis once in 24 hours. Outside this there was nothing, or a sort of unformed chaos. The chief problem of antiquity was to work out how the bodies between the earth and stars moved. These were the moon and sun, and the planets, Mercury, Venus, Mars, Jupiter and Saturn. These rose and set, but not in exact time with the stars. The moon rises about an hour later every day, the sun about four minutes (reckoned by the stars). The planets behave very oddly, for although they generally very slowly fall behind the stars, they sometimes stay still relatively to them and even gain on them for a short time. The Greeks decided that the only *fitting* path for a heavenly body was the 'perfect'

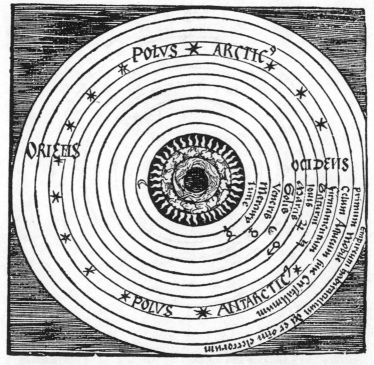

FIG. 4.—The Aristotelian scheme of the universe, as portrayed in the middle ages. From within, outwards, (1) Spheres of earth, water, air and fire: these compose the terrestrial region; (2) the planetary spheres carrying Moon, Mercury, Venus, Sun, Mars, Jupiter, Saturn and the fixed stars. Then come (3) the crystalline sphere which provided the rotation of the pole, needed for the precession of the equinoxes (unknown to Aristotle); (4) the first mover, which impels the rest, and (5) the habitation of God and the Saints.

figure of a circle; and Aristotle, indeed, laid down that just as on earth bodies, if undisturbed, naturally fell down (i.e. towards the centre), so heavenly bodies *naturally* moved uniformly in circles round the centre.

The problem the Greeks set themselves was to discover a combination of circular motions which would give rise to the curious paths which the planets in fact took. The first important solution was that of concentric spheres. Imagine a 'nest' of spheres one inside the other, turning on axles like

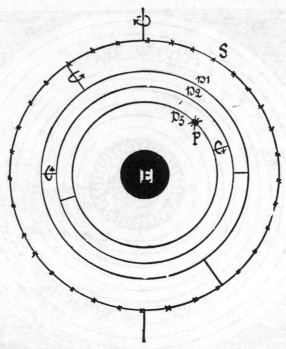

FIG. 5.—Eudoxian or concentric system of planetary motion, later modified by Callippus and adopted by Aristotle. The figure shows the scheme for one planet only, e.g. Saturn. The Earth E is stationary at the centre. The starry sphere S rotates about the centre of the earth: in this sphere and carried about with it are the axes of a second sphere D1 rotating at a different rate: in this arc are the axes of a third sphere D2 which carries a fourth D3. To this is attached the planet whose motion is therefore a combination of the rotations of all four spheres. (From *Blackfriars Monthly Review*, by courtesy of the Editor.)

wheels (Fig. 5). Each sphere has its axle set in the surface of the sphere next larger than itself. The axles incline in various directions and the spheres rotate at very different speeds. The planet, borne on the innermost sphere, will combine the motions of all of them, and will describe at intervals a sort of figure-of-eight path which is not unlike its apparent path through the heavens. The correspondence of this system with the observed facts was not at all exact; and another difficulty was that the planets change in brightness and so seem sometimes to be nearer to the earth and sometimes further from it, which could not be the case if they moved on a sphere concentric with it.

The idea of concentric invisible spheres, adopted though not invented by Aristotle, persisted until the seventeenth century; but at the same time another and better system, the Ptolemaic, was used by astronomers from the second to the seventeenth

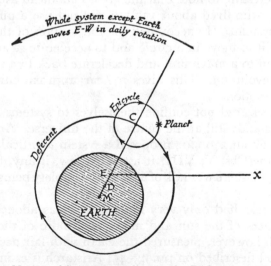

FIG. 6.—Motion of Saturn according to the Ptolemaic system. The Earth is stationary and its centre M is the centre of the universe. The planet travels in the epicycle, whose centre travels round the deferent circle whose centre is at D. It does not move uniformly about D, but about E the centre of the equant circle: i.e. so that the angle CEX increases uniformly. Meanwhile the whole system is carried about the earth once a day by the motion of the heavens.

century. The earth, as before, is supposed to be motionless at the centre and the planets to move round it. Each planet moves round a circle, the epicycle, the centre of which moves in another circle which surrounds the earth, though its centre is not at the earth's centre. The centre of the epicycle does not, however, move uniformly but so as to lie always on the uniformly rotating radius of a third circle. This complicated system, if the size of the circles, the position of their centres and the rate of rotation were properly chosen, could give a very fair account of the motion of the planets. This was important in all ages up till about 1700, because men took much account of astrology, which foretold the events of their lives by the relative positions of the stars and planets. Although by modern standards these systems were far from accurate, they were a great advance on no system at all, and, moreover, were wonderful feats of mathematical reasoning.

It is interesting to note that the great Chaldæan astronomer, Kiddinnu, who lived about 150 B.C. did not use a purely geometrical system. He accounted for the motion of the sun by supposing it to move in a circle and to accelerate steadily from a minimum to a maximum and decelerate back to a minimum in each revolution. This gives a near approximation to its apparent motion.

The Greeks did not confine themselves to systems in which the earth stands still at the centre of the universe. Aristarchus of Samos, about 270 B.C., proposed a system identical with the Copernican (Chs. V, VII); it attracted few, if any, followers, however, and there was talk of a charge of impiety being brought against him.

The Greeks had only very inaccurate ideas about the sizes and distances of the sun and moon, still more of the planets. They had, however, measured the earth with fair accuracy by the method described on pp. 43–45; Aristarchus estimated the distances and sizes of the sun and moon by a method which was sound in theory and, although it gave very inaccurate results in his hands, served to demonstrate that the sun was many times bigger than the earth and much further away than the moon.

Greek Mathematics

The modern child starts with arithmetic and goes on to algebra and geometry. The Greeks made very little study of arithmetic, had almost no algebra, but did practically everything in geometry that can be done without algebraical or trigonometrical help. Euclid, about 300 B.C., collected and systematised the geometry of his time, and seventy or eighty years ago a literal translation of Euclid's *Elements* was the standard text-book of geometry. Euclid worked up through his thirteen books to the construction of the five regular solids (p. 39) by the use of ruler and compass only. Later Greeks went much further than Euclid and worked out the geometry of the figures we get by slicing a cone (the ellipse, parabola and hyperbola), and invented many other special curves. They worked out the geometry of spheres very carefully, because they needed it for astronomy. They made very great efforts to find geometrical constructions to solve three problems that cannot in fact be solved in that way; squaring the circle (i.e. constructing two straight lines whose ratios are $1 : \pi$); constructing a cube of double the volume of a given cube (this involves finding $\sqrt[3]{2}$); and trisecting an angle. A great deal of valuable geometry was discovered in their efforts: and ways of *approximating* to solutions of all three were discovered (pp. 40–42).

Greek arithmetic could not get very far because their system of numeration was like the Roman and did not depend on position: calculation was possible but laborious. Nothing like algebra is found till near the close of Greek activity, and then may have come from India, where arithmetic and algebra developed much quicker than geometry.

Physics

The Greeks made some study of optics, because shadows and rays followed geometrical laws and were important in astronomy, e.g. in predicting eclipses. They worked out the laws of reflection in mirrors, but they did not understand refraction; and though they seem to have had burning-glasses they do not record any study of lenses.

Aristotle greatly hampered physics and astronomy by build-

ing a system on two assumptions which he omitted to check
by experiment. He had evidently noticed that light bodies
fell more slowly than heavy ones (e.g. leaves than stones) and
that bodies fell more swiftly through air than water, and more
swiftly through water than, e.g., honey. So he deduced that
the speed of fall of a body was (1) proportional to its weight,
(2) inversely proportional to the resistance of the medium.
So a ten-pound weight would fall ten times as quickly as a
one-pound weight: and if water had 20 times the resistance
of air things would fall in air 20 times as fast as they would
sink in water. Neither of these rules agree with the facts, and
Aristotle unfortunately used them as axioms. He thus proved
by quite sound reasoning a number of false conclusions (e.g.
that a vacuum could not exist). Consequently mechanics had
to wait nearly two thousand years to make a start.

Archimedes, a very great mathematician, measured densities
and studied centres of gravity and the way things float, but
rather as an exercise in geometry than as physics. He even
constructed mechanical devices and engines of war, though we
are told that he did not consider these material applications

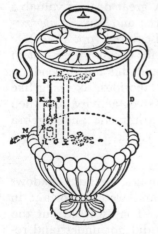

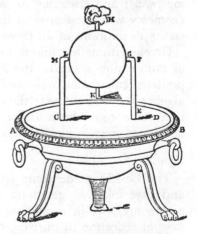

FIG. 7.—"Penny-in-the
slot" machine for de-
livering holy water, as
described by Heron of
Alexandria.

FIG. 8.—Heron's reaction turbine.
Water is boiled by a fire beneath
the cauldron and steam jets issuing
from the pivoted globe cause it to
rotate.

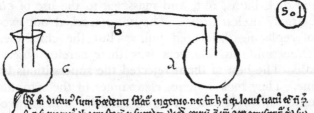

(50.1)

FIG. 9.—XII–XIII century manuscript of Philon of Byzantium showing his instrument for demonstrating the expansion of air by heat. The sphere *a* contains some air and some water into which the central tube dips. The heat of the sun causes the air to expand and drive water through the tube *b* into the flask *c*. (From *The Annals of Science*, Vol. V, No. 2.)

of science as worthy of record. When Greek science was already beginning to decline, two authors—and probably others whose works are lost—began to study the flow of liquids and the properties of gases, rather as a way of making ingenious toys than as real science. Heron of Alexandria discovered the first steam-engine—to him a mere toy, and Philon of Byzantium describes what might have been used as a thermometer (Fig. 9).

The Romans were quite good engineers, they made bridges, pumps, aqueducts, theatrical machinery, engines of war and the like, but they were not interested in the theory of such things. The Greeks were all theory and no practice: the Romans were the opposite, and it was only in the eighteenth and nineteenth centuries that we began effectively to combine these two aspects of science.

Greek and Roman Medicine

Medicine is a science that cannot be entirely theoretical because patients have to be treated. The first Greek medical practice was by the priest-physicians of the temples of Asklepios, god of healing, but in the fifth and sixth centuries there arose a class of lay physicians of whom the most famous was Hippocrates. Their practice was, on the whole, sensible: they preferred diet, baths, rest, and massage, to the use of quantities of largely useless drugs. They also worked out excellent ways of replacing dislocated joints. But the chief merit of Hippocrates and his followers was their careful observation of disease. The best of them rejected the superstitious type of medicine. They took concise, clear notes of the symptoms of illnesses at all stages and set down rules by which the course of the disease might be predicted. In the third century B.C. the Alexandrian surgeons studied anatomy much more carefully, and acquired a fair knowledge of how the body was constructed, although they had scarcely any knowledge of how it worked. After this period medicine seems to have become less scientific. More and more drugs became known, and medicine tended to degenerate into the administration of elaborate messes of ingredients, many of which we should regard to-day as disgusting in the extreme. The main body of medicine was finally summed up by Galen, about A.D. 150, in several long treatises. The work of Galen and Hippocrates was the basis of almost all medicine until the sixteenth century and remained influential until the eighteenth.

Anatomy and Physiology

There was a prejudice against dissecting the human body at most periods of antiquity, though at Alexandria, in the third century B.C., it was allowed, and there are even stories of human vivisection. So even Galen had a good many serious anatomical mistakes, resulting from his deduction of human anatomy from animal dissections.

Of physiology—the way the body worked—there were numerous theories, but almost no understanding; of these theories the following is only a sketch. There were generally supposed to be four body-fluids (humours), blood, phlegm, yellow bile and black bile, the quantities of which were perfectly proportioned in the healthy body. The different temperaments as well as diseases depended on an excess of one of these; thus we still speak of people as *sanguine, phlegmatic, choleric* or *melancholy* in temperament. The physiology of Galen may be taken as typical. Blood was not believed to circulate, but

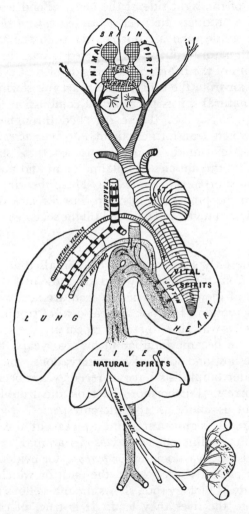

FIG. 10.—The Physiology of Galen (from C. Singer. *Greek Biology and Greek Medicine*). (By courtesy of the Oxford University Press.)

to ebb and flow in the veins as a result of the expansion and contraction of the right side of the heart. Food was digested, and from the products the liver separated *natural spirits*, something more subtle than air but coarser than the soul. Blood charged with natural spirits was supposed to pass through pores in the septum of the heart to the left side of that organ, where it met with air from the lungs. The heart transformed the air, blood and natural spirits by a sort of combustion to the even more subtle *vital spirits*. These travelled through arteries to the brain which transformed them into *animal spirits*, something almost like mind. The brain forced these spirits along the nerves into the muscles so inflating them and making them swell up and shorten (i.e. contract). These theories were of no assistance to the practical physician, but for lack of anything better they were taught and believed till the seventeenth century.

Biology

Almost the only serious studies of animals that have come down to us from the classical world are those of Aristotle. His descriptions of the creatures he has seen are accurate and full of good observation. He took a great deal of trouble to watch the animals he was able to study, especially the sea-creatures with which he became familiar in his two years' stay on the island of Lesbos, and some of his observations which were rejected by later naturalists have in recent years been found to be quite correct. Furthermore he took the trouble to make dissections of as many as 48 different species. Moreover he grasps the really significant points, picks out the really important points of difference and *classifies* animals in a sensible fashion. Aristotle was looking for *purpose*, for evidence that an animal was perfectly adapted to the end for which it strives, and any study of biology soon shows the marvellous adaptation of animals to the lives they lead. It is true, of course, that without the microscope or any knowledge of chemistry, he could not discover much about the inner workings of the body and many of his explanations were incorrect: nevertheless he stands out as immeasurably the greatest biologist who lived before the time of modern science.

His pupil Theophrastus has left a text-book of botany which shows the same admirable observation. It is not so attractively presented as are the works of Aristotle but one can find in it important features of classification of plants and their organs, and nothing like it is found again before the sixteenth century. No one else in antiquity produced any first-class studies of the departments of biology other than medicine. Plants are chiefly studied from the druggists' point of view; and, as time goes on, zoology, started so well by Aristotle, degenerates into uncritical fables about beasts and the moral lessons to be learnt from them. The Greeks were the founders of biology as a science, the study of organisms for their own sake with a view to eliciting general laws concerning them. They set out the main classes of animals, and recorded the main outlines of anatomy. They generally assigned the true function to each organ, though they did not understand how it operated,

The decline of science

The Greeks were great logicians and geometers, and fair astronomers; they made a beginning in many other sciences, but no more than a beginning. They were the originators of the scientific spirit and of almost all the science of classical times; for in this the Romans were only their followers. Their best work was finished by 150 B.C. After that we have plenty of encyclopædists who compile long books from the works of others and may or may not add something themselves. Pliny's *Natural History*, though containing masses of information, is poor uncritical stuff. Only Galen the physician and Ptolemy the astronomer, who date from the second century A.D., are anything more than mere compilers. After about A.D. 300 interest in natural science almost ceases: Christianity, new and living, was giving men what they had always been seeking, and in that new world of inspiration and love, nothing seemed important except to live well and know the Divine truth. So science almost disappears and for centuries the learned world is busy confuting heresies and defining exactly what a Christian can believe without fear of error.

Examples of Greek Science

PLATO ON THE STUDY OF GEOMETRY

Those then who are but a little conversant in geometry, said I, will not dispute with us this point at least, that this science is perfectly contrary to the common modes of speech employed about it by those who practise it. How? said he. They speak very ridiculously, and as if through poverty of ideas; for all the discourse they employ in it appears to be with a view to actual practice. Thus they speak of making a square, of prolonging, of adjoining, and the like. But yet the whole of this discipline is studied for the sake of knowledge. By all means, said he. Must not this further be assented to? What? That it is the knowledge of that which always is, and not of that which is sometimes generated and destroyed. This, said he, must be granted; for geometrical knowledge is of that which always is.

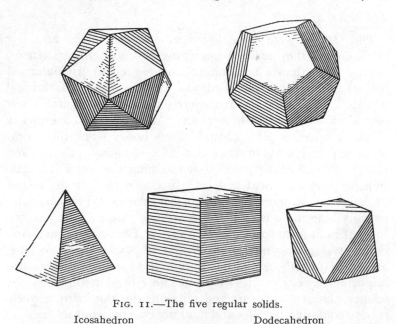

FIG. 11.—The five regular solids.

Icosahedron		Dodecahedron
Tetrahedron	Cube	Octahedron

It would seem then, generous Glauco, to draw the soul towards truth, and to be productive of an intellectual energy adapted to a philosopher, so as to raise this power of the soul to things above, instead of causing it improperly, as at present, to contemplate things below.

(From *The Republic of Plato*, translated by Thomas Taylor. Book VII.)

A PROPOSITION OF EUCLID, BOOK XIII

I say next that *no other figure, beside the said five figures, can be constructed which is contained by equilateral and equiangular figures equal to one another.*

For a solid angle cannot be constructed with two triangles or indeed planes.

With three triangles the angle of the pyramid is constructed, with four the angle of the octahedron, and with five the angle of the icosahedron:

> but a solid angle cannot be formed by six equilateral and
> equiangular triangles placed together at one point,
> for the angle of the equilateral triangle being two-thirds of
> a right angle, the six will be equal to four right angles:
> which is impossible, for any solid angle is contained by
> angles less than four right angles.

For the same reason neither can a solid angle be constructed by more than six plane angles (triangles?).

By three squares the angle of the cube is contained, but by four it is impossible for a solid angle to be contained.

For they will again be four right angles.

By three equilateral and equiangular pentagons the angle of the dodecahedron is contained;

> but by four such it is impossible for any solid angle to be
> contained;
> for, the angle of the equilateral pentagon being a right
> angle and a fifth, the four angles will be greater than four
> right angles:
> which is impossible.

Neither again will a solid angle be contained by other polygonal figures by reason of the same absurdity.

Therefore no other figure beside the said five figures can be constructed which is contained by equilateral and equiangular figures equal to one another;

which was required to be proved.

(*The XIII books of Euclid's Elements.* Sir Thomas L. Heath, C.U.P.)

EXAMPLES OF ATTEMPTS TO SOLVE THREE GREAT PROBLEMS

The squaring of the circle. Euclidean methods are incapable of constructing a square equal to a given circle, but Hippocrates of Chios about 440 B.C. showed how to construct a square equal in area to a *lune*, bounded by two circles.

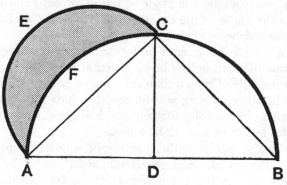

FIG. 12.—Quadrature of a lune.

In Fig. 12 ACB is a semi-circle, on the diameter AB. DC is a radius perpendicular to AB. AEC is a semi-circle on the chord AC as diameter. The student can easily prove that the area of the lune AECF is equal to that of the square on half the line AC.

Trisection of an angle. To trisect the angle AÔD, draw a circle, with centre O, to cut AO in E; produce DO. Draw a straight line EBC so that OB = BC. Then BOC = 1/3 AÔD.

Is this construction sound? Can it be done with ruler and compass only?

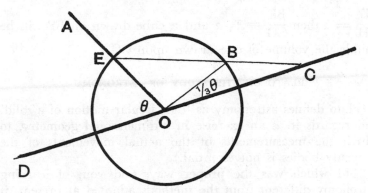

FIG. 13.—Trisection of an angle.

Duplication of the cube

The problem is to construct a cube of volume double that of another cube. If the side of the latter is of length 1, that of the former must be $\sqrt[3]{2}$. The problem is therefore to construct a line of length $\sqrt[3]{2}$. This cannot be done by Euclidean geometry. Eratosthenes adopted the construction shown in Fig. 14. AX, EY are two parallels, 2 units apart, and AMF, MNG, NQH, are three equal set-squares. The perpendicular DH of length 1 unit is drawn. Then the set-squares are adjusted so that their intersections B, C, lie on the line AD. Then

$$\frac{AE}{BF} = \frac{BF}{CG} = \frac{CG}{DH}$$

and if

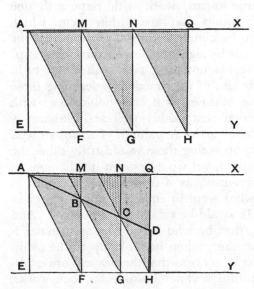

FIG. 14.—The finding of two mean proportionals, leading to the duplication of the cube.

$\dfrac{AE}{DH} = 2$ then $\dfrac{BF}{CG} = {}^3\sqrt{2}$ and a cube drawn on BF will be double the volume of one drawn upon CG.

PLATO ON THE STUDY OF ASTRONOMY

(Plato defines astronomy as 'the circular motion of a solid' and regards it as an exercise in 3-dimensional geometry, to which the measurements of the actual movements of the heavenly bodies is not essential.)

"But which was the proper way, said you, of learning astronomy different from the methods adopted at present, if they mean to learn it with advantage for the purposes we speak of? In this manner, said I; although these variegated bodies in the heavens are deemed the most beautiful and the most accurate of the kind, yet (as they are only part of the *visible* world) are far inferior to the real beings which are carried in those orbits in which real velocity, and real slowness, in true number, and in all true forms, work with respect to one another, and carry all things that are within them: which latter things truly are to be comprehended by reason and the intellectual power, but not by sight: or do you think they can? By no means, replied he. Is not then, said I, that variety in the heavens to be made use of as a model for learning those real things, in the same manner as if one should meet with geometrical figures, drawn remarkably well and elaborately by Dædalus, or some other artist or painter? For a man who was skilled in geometry, on seeing these would truly think the workmanship most excellent, yet would esteem it ridiculous to consider these things seriously, as if from thence he were to learn the truth, as to what were in equal, in duplicate, or in any other proportion. It would be ridiculous, replied he. And do not you then think, that he who is truly an astronomer is affected in the same manner, when he looks up to the orbits of the planets? And that he reckons that the heavens are established in the most beautiful manner possible for such works; but would not he deem him absurd, who should imagine that this proportion of night with day, and of both these to a month, and of a month to a year, and of other stars to the sun and moon and towards one another, existed always in the same manner, and in no way suffered any change, though they have a body and are visible; and who would search by every method to apprehend the truth of these things. So it appears to me, replied he, whilst I am hearing you. Let us then make use of problems, said I, in the study of astronomy, as in geometry. And let us dismiss the heavenly bodies, if we intend truly to apprehend astronomy, and render profitable instead of unprofitable that part of the soul which is naturally wise."

(From *The Republic of Plato*, tr. Thomas Taylor. Book VII.)

MEASUREMENT OF THE EARTH BY ERATOSTHENES

Syene and Alexandria lie, he says, under the same meridian circle. Since meridian circles are great circles in the universe, the circles of the earth which lie under them are necessarily also great circles. Thus, of whatever size this method shows the circle on the earth passing through Syene and Alexandria to be, this will be the size of the great circle of the earth. Now Eratosthenes asserts, and it is the fact, that Syene lies under the summer tropic. Whenever, therefore, the sun, being in the Crab at the summer solstice, is exactly in the middle of the heaven, the gnomons (vertical pointers) of sundials necessarily throw no shadows, the position of the sun above them being exactly vertical; and it is said that this is true throughout a space three hundred stades in diameter. But in Alexandria, at the same hour, the pointers of sundials throw shadows, because Alexandria lies further to the north than Syene. The two cities lying under the same meridian great circle, if we draw an arc from the extremity of the shadow to the base of the pointer of the sundial in Alexandria, the arc will be a segment of a great circle in the (hemispherical) bowl of the sundial, since the bowl of the sundial lies under the great circle (of the meridian). If now we conceive straight lines produced from each of the pointers through the earth, they will meet at the centre of the earth. Since then the sundial at Syene is vertically under the sun, if we conceive a straight line coming

from the sun to the top of the pointer of the sundial, the line
reaching from the sun to the centre of the earth will be one
straight line. If now we conceive another straight line drawn
upwards from the extremity of the shadow of the pointer of
the sundial in Alexandria, through the top of the pointer to
the sun, this straight line and the aforesaid straight line will
be parallel, since they are straight lines coming through from
different parts of the sun to different parts of the earth. On

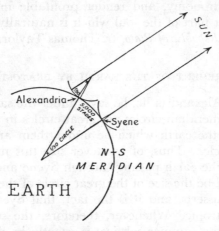

FIG. 15.—Measurement of the circumference of the earth,
according to Eratosthenes.

these straight lines, therefore, which are parallel, there falls
the straight line drawn from the centre of the earth to the
pointer at Alexandria, so that the alternate angles which it
makes are equal. One of these angles is that formed at the
centre of the earth, at the intersection of the straight lines
which were drawn from the sundials to the centre of the earth;
the other is at the point of intersection of the top of the pointer
at Alexandria and the straight line drawn from the extremity
of its shadow to the sun through the point (the top) where it
meets the pointer. Now on this latter angle stands the arc
carried round from the extremity of the shadow of the pointer
to its base, while on the angle at the centre of the earth stands
the arc reaching from Syene to Alexandria. But the arcs are

similar, since they stand on equal angles. Whatever ratio, therefore, the arc in the bowl of the sundial has to its proper circle, the arc reaching from Syene to Alexandria has that ratio to *its* proper circle. But the arc in the bowl is found to be one-fiftieth of its proper circle. Therefore the distance from Syene to Alexandria must necessarily be one-fiftieth part of the great circle of the earth. And the said distance is 5000 stades; therefore the complete great circle measures 250,000 stades.

(40 stades = 12,000 royal Egyptian cubits of 52.5 cm.: therefore 250,000 stades is 24,662 miles. The true circumference is about 24,900 miles. The angle of the shadow is correct within a minute or so of arc, but Alexandria is by no means directly north of Syene (the modern Aswan): it is not clear whether Eratosthenes meant 5000 stades to be the distance between the towns or the distance between their parallels of latitude. If he intended the latter, the measurement is remarkably correct.)

(Translation from *Greek Astronomy*, by Sir Thomas Heath; by courtesy of Messrs. J. M. Dent and Sons.)

ARISTOTLE ON THE CETACEA

The dolphin, the whale, and all the rest of the Cetacea, all, that is to say, that are provided with a blow-hole instead of gills, are viviparous. That is to say, no one of all these fishes is ever seen to be supplied with eggs, but with an embryo from whose differentiation comes the fish, just as in the case of mankind and the viviparous quadrupeds.

The dolphin bears one at a time generally, but occasionally two. The porpoise in this respect resembles the dolphin, and, by the way, it is in form like a little dolphin, and is found in the Euxine; it differs however from the dolphin in being less in size and broader in the back; its colour is leaden-black. Many people are of opinion that the porpoise is a variety of the dolphin.

All creatures that have a blow-hole respire and inspire, for they are provided with lungs. The dolphin has been seen asleep with his nose above water, and when asleep he snores.

The dolphin and the porpoise are provided with milk and

suckle their young. . . . Its young accompany it for a considerable period; and in fact the creature is remarkable for the strength of its parental affection. It lives for many years; some are known to have lived for more than twenty-five and some for thirty years; the fact is fishermen nick their tails sometimes and set them adrift again and by this expedient their ages are obtained.

(*History of Animals. Aristotle.* Ed. Smith & Ross. Trans. D'Arcy Wentworth Thompson. Oxford, 1910. 566^b. 12. 1–27.)

THE STUDY OF BIOLOGY

Having laid this foundation, we proceed to the next topic, and by way of introduction we observe that (Chap. VI) some members of the universe are ungenerated, imperishable, and eternal, while others are subject to generation and decay. The former* are excellent beyond compare and divine, but less accessible to knowledge. The evidence that might throw light on them, and on the problems which we long to solve respecting them, is furnished but scantily by sensation; whereas respecting perishable plants and animals we have abundant information, living as we do in their midst, and ample data may be collected concerning all their various kinds, if only we are willing to take sufficient pains. Both departments, however, have their special charm. The scanty conceptions to which we can attain of celestial things give us, from their excellence, more pleasure than all our knowledge of the world in which we live; just as a half-glimpse of persons that we love is more delightful than a leisurely view of other things, whatever their number and dimensions. On the other hand in certitude and in completeness our knowledge of terrestrial things has the advantage. Moreover, their greater nearness and affinity to us balances somewhat the loftier interest of the heavenly things that are the objects of the higher philosophy. Having already treated of the celestial world, as far as our conjectures could reach, we proceed to treat of animals, without omitting to the best of

*The heavenly bodies.

our ability, any member of the kingdom, however ignoble. For if some have no graces to charm the sense, yet even these, by disclosing to intellectual perception the artistic spirit that designed them, give immense pleasure to all who can trace links of causation, and are inclined to philosophy. Indeed, it would be strange if mimic representations of them were attractive, because they disclose the mimetic skill of the painter or sculptor, and the original realities themselves were not more interesting, to all at any rate who have eyes to discern the reason that presided over their formation. We therefore must not recoil with childish aversion from the examination of the humbler animals. Every realm of nature is marvellous: and as Heracleitus, when the strangers who came to visit him found him warming himself at the furnace in the kitchen, is reported to have bidden them not to be afraid to enter, as even in that kitchen divinities were present, so we should venture on the study of every kind of animal without distaste; for each and all will reveal to us something natural and something beautiful. Absence of haphazard and conduciveness of everything to an end are to be found in Nature's works in the highest degree, and the resultant end of her generations and combinations is a form of the beautiful.

(From *On the Parts of Animals*. *Aristotle*. Trans. by
W. Ogle, 1882. Kegan, Paul, Trench & Co.)

THEOPHRASTUS ON THE GERMINATION OF SEEDS

Some seeds in germinating put forth their primary root and leaf from one and the same point; others, the root from one end and the leaf from another. . . . Wheat, barley, rye, and all the grains sprout from both ends; that is to say the basal and thicker end of the grain puts forth the root, the upper and narrower end the green herbage. The two, however, are connected and continuous as one. But neither the bean nor any seeds of leguminous plants have this way of sprouting. These put forth root and stem from the same part, namely, that at which the seed was linked to the pod, as if under that point lay the special seat of the growing principle. In the case of seeds of this kind the root at first appearing begins to show a downward tendency, the stem an upward.

(Theophrastus. *Natural History of Plants.* Translated by E. L. Green in his *Landmarks of Botanical History*, pp. 95-96. Smithsonian Miscellaneous Collections, Vol. 54.)

HIPPOCRATES ON THE SACRED DISEASE

"I am about to discuss the disease called 'sacred'.* It is not, in my opinion, any more divine or more sacred than other diseases, but has a natural cause, and its supposed divine origin is due to men's inexperience and to their wonder at its peculiar character. Now while men continue to believe in its divine origin because they are at a loss to understand it, they really disprove its divinity by the facile method of healing which they adopt, consisting as it does of purifications and incantations. But if it is to be considered divine just because it is wonderful, there will be not one sacred disease but many, for I will show that other diseases are no less wonderful and portentous, and yet nobody considers them sacred. For instance quotidian fevers, tertians and quartans, seem to me to be no less sacred and god-sent than this disease, but nobody wonders at them. Then again one can see men who are mad and delirious from no obvious cause, and committing many strange acts, while in their sleep, to my knowledge, men groan and shriek, others choke, others dart up and rush out of doors, being delirious until they wake, when they become as healthy and rational as they were before, though pale and weak; and this happens not once but many times. Many other instances, of various kinds, could be given, but times does not permit us to speak of each separately."

"But this disease in my opinion is no more divine than any other; it has the same nature as other diseases, and the cause that gives rise to individual diseases. It is also curable, no less than other illnesses, unless by long lapse of time it be so ingrained as to be more powerful than the remedies that are applied. Its origin, like that of other diseases, lies in heredity.

*Epilepsy

For if a phlegmatic parent has a phlegmatic child, a bilious parent a bilious child, a consumptive parent a consumptive child, and a splenetic parent a splenetic child, there is nothing to prevent some of the children suffering from this disease when one or other of the parents suffered from it; for the seed comes from every part of the body, healthy seed from the healthy parts, diseased seed from the diseased parts. Another strong proof that this disease is no more divine than any other is that it affects the naturally phlegmatic, but does not attack the bilious. Yet if it were more divine than others, this disease ought to have attacked all equally, without making any difference between bilious and phlegmatic.

The fact is that the cause of this affection, as of the more serious diseases generally, is the brain."

(*Hippocrates*, Vol. II, p. 139 ff. N. W. H. Jones. Loeb Library. Heinemann, London.)

CASE DESCRIPTION FROM HIPPOCRATES

In Thasos the wife of Delearces, who lay sick on the plain, was seized after a grief with an acute fever with shivering. From the beginning she would wrap herself up, and throughout, without speaking a word, she would fumble, pluck, scratch, pick hairs, weep and then laugh, but she did not sleep; though stimulated, the bowels passed nothing. She drank a little when the attendants suggested it. Urine thin and scanty; fever slight to the touch; coldness of the extremities.

Ninth day. Much wandering followed by return of reason; silent.

Fourteenth day. Respiration rare and large with long intervals, becoming afterwards short.

Seventeenth day. Bowels under a stimulus passed disordered matters, then her very drink passed unchanged; nothing coagulated. The patient noticed nothing; the skin tense and dry.

Twentieth day. Much rambling followed by recovery of reason; speechless; respiration short.

Twenty-first day. Death.

The respiration of this patient throughout was rare and large;

took no notice of anything; she constantly wrapped herself up; either much rambling or silence throughout.

(From the *Works of Hippocrates*, Vol. I, Epidemics III, Case XV. Translated by W. H. S. Jones, Loeb Library.)

HIPPOCRATES ON BATHS

The bath will be beneficial to many patients, sometimes when used continuously, sometimes at intervals. Occasionally its use must be restricted, because the patients have not the necessary accommodation, for few houses have suitable apparatus and attendants to manage the bath properly. Now if the bath be not carried out thoroughly well, no little harm will be done. The necessary things include a covered place free from smoke, and an abundant supply of water permitting bathings that are frequent but not violent, unless violence is necessary. If rubbing with soap be avoided, so much the better; but if the patient be rubbed, let it be with soap that is warm, and many times greater in amount than is usual, while an abundant affusion should be used both at the time and immediately afterwards. A further necessity is that the passage to the basin should be short, and that the basin should be easy to enter and to leave. The bather must be quiet and silent; he should do nothing for himself but leave the pouring of water and the rubbing to others. Prepare a copious supply of tepid water, and let the affusions be rapidly made. Use sponges instead of a scraper, and anoint the body before it is quite dry. The head, however, should be rubbed with a sponge until it is as dry as possible. Keep chill from the extremities and the head, as well as from the body generally. The bath must not be given soon after gruel or drink has been taken, nor must these be taken soon after a bath.

(*Works of Hippocrates*. Translated by W. H. S. Jones. Loeb Library. Regimen in acute diseases, LXV.)

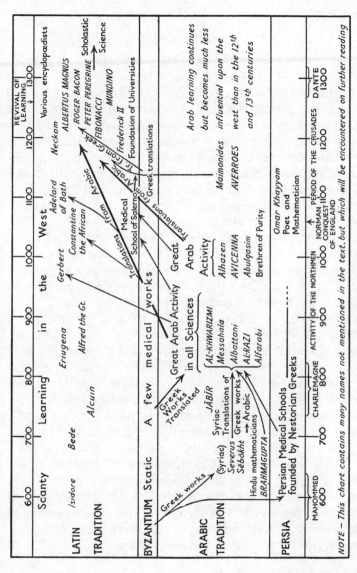

FIG. 16.—Science in the Middle Ages

CHAPTER FOUR

Science in Eclipse

The period of tradition

The great wave of Greek Science began to subside after about 200 B.C., and after about 150 B.C. not much original work was done. A highly cultured and organised society, skilled in the fine arts and practical crafts, persisted for some five hundred years, during which the works of the great Greek mathematicians, astronomers and physicians remained well known, but little was added to them except comment and explanations. At this time there were three great centres of culture. The first was Alexandria, where an extraordinary medley of races, Egyptians, Chaldæans, Phœnicians, Jews, Greeks, Persians, and the like possessed in common the Greek language and culture. It remained a vigorous centre of learning from 306 B.C. to A.D. 642, when it was taken by the Moslems.

Rome was less a centre of science than of literature and history and law; its language was Latin, though Greek was widely read for a time. As a centre of culture it greatly declined after its sack by Alaric in 410, and when Roman knowledge of Greek died out in the fifth century, so did most of the ancient science, very little of which had been translated into Latin.

Byzantium was not a great centre of learning before A.D. 330, but after that date, when Constantine refounded it as the seat of the Eastern Empire, it grew in importance, and remained a city of high though sterile Greek culture till its sack by the Turks in 1453. It was in Byzantium that the classical Greek tongue continued to be read and Greek manuscripts of ancient authors to be preserved; and she handed on this knowledge of Greek learning, first to the Syrians and Persians, thence to the Moslem world, and later to Western Europe.

The beginning of Chemistry

Chemistry, in the sense of something that is done with special chemical apparatus in a laboratory, appears first in Alexandria about A.D. 100 in the form of Alchemy, which purported to be the art of transforming other metals into gold—a process which was quite credible on the ancient theory of matter. Alchemy throve for fifteen hundred years, and though we may suppose that the alchemists made no gold, they invented chemical technique. In their works we first hear of distillation, the inventor of which seems to have been a woman, Maria the

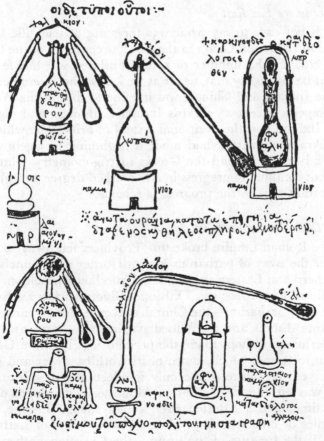

FIG. 17.—Distillation apparatus as used by the Greek alchemists, c. A.D. 100–300.

Jewess, who lived in the first or second century A.D. They used flasks, beakers, funnels, filters, water-baths, ash-baths, condensers, receivers, sublimation apparatus, and until the nineteenth century very little was added to the apparatus they invented. The alchemists were primarily interested in the problem of making gold, but incidentally discovered a number of new materials; though we cannot call the harvest of 1500 years of their investigations a large one.

Science in the Far East

It is not easy to say what was the state of scientific knowledge in India and China in the first few centuries of the Christian era, for the number of texts translated are but few. It might be fair to say that, as far as we know, the *scientific* culture of the Indians and Chinese was inferior to that of the West in all respects except as regards Indian arithmetic and algebra. The Indians had the positional system of arithmetic which we call Arabic numerals, had made a beginning in algebra, and had advanced beyond the Greeks in trigonometry. But they did not continue to progress in any marked degree and play no important part in the progress of science.

The Dark Ages in the West

The Roman Empire broke up. Provinces fell away to come under the sway of barbarous kings. Hordes of the uncivilised Northern and Eastern tribes poured into lands which once had a polished Roman society. Only one influence preserved Europe from utter barbarism—the Church of Rome. Missionaries went out into distant and uncivilised pagan lands, such as Britain or Germany; bishops made the perilous journey from Gaul or Spain to Rome, which, shorn of its worldly power and glory, remained the centre of the only intellectual and spiritual life that was left, the life of religion. Monasteries developed during this time; and so, dotted over a half savage Europe, were colonies of men who kept alive the knowledge of Latin and with it the treasure of the liturgy, the Christian Fathers, and the Scriptures. Between, say, A.D. 500 and A.D. 1000 no one in

Western Europe knew much about anything except religion, and the knowledge of scientific matters was particularly deficient. The learned monk would know how to do simple arithmetic in Roman figures, and could give a bare outline of some part of the traditional Greek astronomy; he would, however, have a good knowledge of the theory of the calendar, and, if he was inclined to the study, a certain amount of traditional medical science, partly folk-knowledge, partly traditions of Greek medicine, partly white-magic or superstition. His book-knowledge of plants and animals was chiefly from the latest and worst Greek tradition of moral stories about beasts and plants. But the study of science was not the primary purpose of a monk; rather was he concerned to set up and maintain some standard of religion and decency among savage, treacherous, and ignorant barbarians.

Culture in the East

Alexandrian scientific learning survived in a rather unpractical and scholarly form for many centuries. Much, if not all, of its knowledge was in the possession of the scholars of Byzantium, so when the followers of Mahommed swept over Egypt in the seventh century, and finally conquered all North Africa, Sicily, most of Spain and much of Syria and the Middle East, Greek learning was not lost. The Moslems at first showed a natural intolerance of 'infidel' learning, but within a century they were already gaining a respect for it, and soon were most anxious to gain for themselves all its treasures. Despite this respect, most of the great men who wrote in Arabic were not Arabs, but Nestorian Christians, Jews, or Persians. It seems then that between A.D. 500 and 600 many of the important Greek works were translated into Syriac, and were later translated into Arabic, so that by A.D. 850 the Arabs possessed translations of Aristotle, Euclid, Ptolemy, Galen and many other authors. Their level of culture in subsequent centuries rose far above that of Europe. They had universities, hospitals, fine libraries, observatories, large and flourishing cities with all the refinements of luxury; and they regarded the 'Franks' with no little contempt. Finally the Arab culture produced a number of great encyclopædists. Such men as al-Rāzi (Rhazes) and Ibn Sinā (Avicenna) wrote on almost every department of human thought. It may be their example that led to the writing of so many encyclopædic works on the sciences in Europe,—an unfortunate result, for he who seeks to know all that is known has little time to advance knowledge. Chemistry—both pharmacy and alchemy—was a favourite study, and, although the greater part of Arab chemistry remains untranslated, we know enough to see that they prepared many substances, e.g. borax, sal-ammoniac, and the mineral acids, with which we suppose the Greeks to have been unfamiliar.

Arab culture in general followed the lines of the Greek, though in some fields, such as mathematics, going considerably beyond it. Their mathematics seem to have derived from other sources also. Only one Greek author, Diophantos, wrote on what we now call algebra, and he may well have taken his ideas from some eastern source. The Babylonians knew the principle of solving simple and quadratic equations, and it is quite likely that this knowledge reached Islam by way of the Sabæans who dwelt in the northern part of Syria and preserved much of the Babylonian culture. The Arabs also received mathematical knowledge from India, whence came what we call the Arabic numerals, and the use of sines in trigonometry. The most famous Islamic mathematician was Mohammed ibn Musa al-Khwārizmi (c. 825). He wrote a book Hisāb al-Diābrwa'l-Mukābala concerned with algebraic calculations and the word 'algebra' is derived from the second word of its title. The Arab mathematicians were the teachers of Western Europe up to the sixteenth century.

Arabic astronomy was more practical than the Greek. A great many stars were named and catalogued; they were better instrument-makers than the Greeks and more accurate observers, but they made no advance in principles over Ptolemy, whose work, known to them as the *Almagest* (greatest of books) was the foundation of all their theory.

Islamic geographers, of course, ranged much more widely than the Greek, but they still clung to the errors of Ptolemy

lnud animal eft mulo fluuio q̃o
dicuntur hydrus. phiſiologus dicit
de eo quoo ſenſis eſt hoc animal inimicū
cocodrillo. & hanc habet natinam ꝓ con
ſuetudinem. Eum udet cocodrillum
in litroribus fluminum oormientem aꝑto
ore uadit ꞇ inuoluit ſe in lunum luti q̃o
poſſit facilius illabi in faucibus eius. Eo
codrillus igitur oe ſubito excitatur ujuũ
tranſglutit eum. Ille autem dilanians
omnia uiſcera eius exit uiuus. exinſcen
bit eius. Sic ergo mors & infernus figurā
habent cocodrilli. qui inimicus eſt domi

dūm urtutū. ꞌꝉ īn monte omnīū fēcꝝ. Panthera ē. beſtia. mīnūtꝭ orbiculꝭ ſuppic
ta. ꞁta ut̄ oculatꝭ ex fuluo circulꝭ mḡ ut̄ alba diſtīnguatͤ uarietate. hꝫ ſemel
omnino parturit. cuí cauſe ꞁratio manifeſta ē. Hā ē ꞁn uͤo marꝭ coaluere tres
catulꝭ matꝭ ad naſcendū turibꝭ pollent. odiunt tēpoꝝ moꝛaſ. Itaꝗ onerata fe
tibꝭ uulua tanꝗ obſtante partū unguibꝭ lacerant. effundit illa partū. ſeu
pocuͤl dimittīt. dolore cogente. Jta poſtea corruptaſ ꞁ ꞁcicatriſꝭ ſedibꝭ. genitale
ſem̄ īn fuſū ꞁi haeret accep̄u.ſꝝ irritū reſilīt ſonat. Hā plꝰmi diē aīalia ē ꞁ acuut̄
unguibꝭ frēqnͭ parere n̄ poſſe. uicīant̄ eīm ꞁconſer ſe mouentibꝭ catulis.

Est animal que dr̄ antalops acerrīmū nimiſ. ꞁta ut̄ nec uenatoꝛ ei
poſſit appropinquare. hꝫ aūt longa cornua ſerre figurā habentia.
ꞁta ut̄ poſſit reſecare arboreſ altaſ ꞁ magnaſ ꞁ ad trā pſternere. cū aūt ti
tierit uadit ad ſtrū magnū eufratem. eſt aūt ibi frutex qui grece dr̄ heri
cine hꝫ uꝛgulta ſubtilia c̄ꝑlixa. ꞁ incipit ludere cornu ad heri cīa. ꞁ dū
ludit obligat cornua ꞁn uꝛgultis eiꝰ. ꞁ dū pugnant ē ſe lībare n̄ poteſt
exclamat uoce magnā. Audienſ uenatoꝛ uoce eiꝰ. uenio c̄ occidit tum.
Sic ꞁ tu ho qui ſtudeſ ſobriꝰ ꞁcaſtͤ ē. ꞁ ſpualitͤ uīuͤe.ꝰ duo cornua ſē duo
teſtamͤta p que poteſ reſecare ꞁexcide oīa uitia corporalia ꞁ ſpualia. Ca
ue ab ebrietate ne obligiſ luxuria ꞁ uoluptate. ꞁ inficiariſ a chabolo. Ymū
eīm ꞁ mulieres apoſtatare faciunt hominem â dō.

PLATE IV

Natural history about A.D. 1150. The upper drawing shows the
antelope. who can only be caught by causing him to entangle his
horns in a tree. The lower drawing shows the unicorn or rhinoceros
who can be tamed and subdued by a virgin. (British Museum,
Add. MS. 11283. f. 3. v.)

in many matters, such as the form of India, of which they could have had accurate knowledge.

Arabic physics was a considerable advance on that of the Greeks, as we know it,—though much of their work has been lost. The excellence of their craftsmanship and their mathematical aptitude stood them in good stead. Al-Biruni, the great geographer, did some accurate and delicate work on specific gravities, which was not equalled before the time of

Uando remanet apud eradicatione detis q
fragit aliqd:tuc op3 vt ponas sup locu cotu3
cu butyro dievno:aut duob°dieb°:doec mol
lificet loc°: dein itromitte ad ip3 algesti.aut
forcipes:quozum extremitates assimilantur
ori auis que nominatur ciconia. forma forcipum.

Sunt extremitates eo2 facte sicut lima de itus:aut sicut
aliskilsegi. Si g no obedit tibi ad exitu cu istis forcipib°:
tunc op3:vt caues sup radice:7 detegas carne tota:7 itro/
mittas istrm cp noiat.i. palaca atali parua.cui° for² e h.

Sit breuis extremitatis grosse paruper:7 no sit ibibitu3
vt no frag.t.si g egredit radix p illud est bonu:7 si no:iu/
nare cu istis istrumentis alys quo2 forme sut iste.
Forma pmi triagulate extreitatis in q e qda grossitudo.

Forma triangulate extremitatis subtilis.

Etia iuuat cu h istro brite duos ramos.cui° bec e for².

FIG. 18.—Surgical instruments figured by
Albucasis, from a latin translation of his
treatise printed in 1500.

Galileo. Ibn al-Haitham, known as Alhazen, published a work *The Treasury of Optics* which was very influential in Western Europe and was not improved on until the sixteenth century.

Arab medicine and biology show acute observation and report many phenomena that the Greeks did not observe. Medicine was hampered however by their religious objection to dissection, without which no great progress could be made. The Arab surgeons were skilled operators and invented a number of instruments.

The Revival in the West

From the time of Charlemagne (c. A.D. 800) we must think of the great nations of Europe taking shape gradually and acquiring more stable government and structure. More and more religious houses were founded, and as there came to be more learned men and more contact between them, so enthusiasm for learning grew. The first centres of learning were the monasteries; but the true function of a monk is not to teach but to pray, and a great step towards universal knowledge was taken when the first Western universities were founded, from about A.D. 1000. Much the earliest was the medical school of Salerno in Italy, but the most influential was the University of Paris, which began near 1100 as a school of logic and was recognised as a University about 1150-70. Oxford dates from before 1200 and Cambridge from a little after.

The twelfth and thirteenth century saw a great revival of learning and produced some of the world's greatest men and books. Its strength was in philosophy rather than in science, but even here a level was reached which was not surpassed in seven centuries before the or centuries after.

The keynote of the revival was the rediscovery of some part of Greek learning, most of which had long been unknown to the Western World. The European scholars came to know that the Arabs possessed these treasures. At places where the Arab, Greek, Jewish and Latin culture met, e.g. in Sicily and Spain, Latin translations of Greek texts, some directly from Greek but more from Arabic versions, began to be made in the twelfth century; and early in the thirteenth century Europe possessed a great part of Greek learning in a form somewhat ruffled by translation from Greek to Syriac, and thence to Arabic and Latin (sometimes perhaps with Spanish as a further intermediary). There were very few Western Europeans who knew Greek, but by these a small number of translations were made from Byzantine MSS. direct into Latin. The most important things that were recovered were the works of Aristotle, of which but a fragment was previously known. Here was a wonderful philosophy and system of knowledge; and at the same time Christianity provided a perfect system of doctrine. It was the task of the thirteenth century, especially of St. Albert (Albertus Magnus) and St. Thomas Aquinas his pupil, to harmonise them into a noble system of Christian philosophy.

Not only the philosophical, but also the scientific, works of Aristotle were studied, as were those of Ptolemy, Galen, Euclid, and the great Arabs such as Avicenna; none the less the men of the age were not really interested in practical science. A good start towards discovering new facts was made in the thirteenth century by three original geniuses, St. Albert, Peter Peregrine, and Roger Bacon, but the men of the later Middle Ages preferred to study what the earlier authors had said rather than to use their own hands and eyes.

Albertus Magnus

St. Albert was the best sort of South German, thorough, painstaking, enormously industrious. He commented on the greater part of Aristotle's biology, and he enriched it with his own knowledge and observations, which he acquired in tramping over Europe on the business of the Church (he was known as the Bishop with the Boots). He was the first naturalist since Aristotle to study insects: he describes whales, mentions the polar-bear and nearly all the German birds and mammals. He seems to have made practical experiments in alchemy. He had a gift for rejecting the fabulous, and above all he understood that Nature is not something to be read of in books, but to be seen with one's own eyes. St. Albert and St. Thomas Aquinas made it clear that learning about other than directly spiritual matters was desirable. Earlier Saints and theologians had been doubtful of this, for they had considered that the only know-ledge worth having was that which showed the way to heaven; but the great men of the thirteenth century saw that *knowledge was a whole* and that all of it tended to the glory and knowledge of God. It is an illusion to suppose that the Church opposed science in the Middle Ages, for almost all the men of science were clerics. The truth is that but very few of the men of the time were interested in science, and that nearly all were ex-tremely concerned with religion, which was the centre of their lives. The only sciences that they felt they needed were medi-cine and astronomy. Even medicine was not very important to them, because they believed that to die well was better than to live badly; and the chief interest of astronomy was its rela-tion to astrology—always in favour when life, limb and fortune are uncertain.

Roger Bacon was very far from being as great in the human sense as St. Albert or St. Thomas. He had an unfortunate boastful and abusive manner that made him enemies, but he seems to have come very near to being an experimental scientist of the modern type. He evidently had a laboratory at Oxford and studied lenses and mirrors with great care. There are in-dications (though no proof) that he had a telescope (p. 65). There is a tradition that he had a telescope and a burning-glass in his study, and that the University ordered them to be destroyed because the students wasted their time in looking through the telescope, and lighting candles with the burning-glass. In any event, he lays the strongest emphasis on the need to perform experiments, and on the value of mathematics in interpreting nature—ideas that did not come into their own until more than three centuries had gone by.

There were other experimentalists in this age. Peter Pere-grine, a Picard who flourished about 1270, wrote a remarkable little treatise on the magnet and compass, but little else has survived (p. 64).

The Mediæval View of the World

The general scheme of the world in the years 1200–1500 was based on that of Aristotle. The universe was a sphere within which revolved invisible spheres carrying the stars, sun, moon and planets (Fig. 4). These were the changeless heavens; within the sphere of the moon was the terrestrial and corruptible region. The earth was a globe motionless at the centre of the spherical universe. Some, such as Roger Bacon, realised that this spherical earth could be circumnavigated. Asia, Europe, some of Africa had been mapped; but America, the Pacific and Australia were unknown.

Everything was made up of matter and form (p. 25). The simplest kinds of stuff were earth, air, fire and water, which could be transformed one into another, but had as a basis a simple prime matter which could never be obtained from them. Every kind of matter, if it could be transformed into this prime matter, could be made into any other kind of matter, but in practice this could not be done or only in a few cases.

Ordinary matter was not living. Plants had vegetative souls (i.e. life); animals had vegetative and sensitive souls; man vegetative, sensitive and rational souls. The heavenly bodies were generally considered to be living things, though this was disputed.

The anatomy of the human body was not nearly so well known as in the time of the Greeks. Dissection was not forbidden, but it was done in a perfunctory way, simply in order to show that the outlines of Galen's teaching were correct.

A notion, really fundamental to the mediæval world, but less important in that of the earlier Greeks, was the idea of *influences*. God exerted a continuous influence upon his creatures on earth, sometimes directly, but chiefly through the heavenly bodies, and above all the sun. Every stone, plant, animal, organ, human being, or nation, was linked with various heavenly bodies and subject to continuous influence from them. As these heavenly bodies moved through the sky these influences were modified. Plagues and pestilences were caused by these influences—hence our word *influenza*. Man was subject to these influences, but not fatalistically bound by them. He made his life through his own free-will, but he had to cope with the evil influences of the planets as best he could, just as he had to cope with drought and heat and cold and rain.

This general system of the world suffered singularly little change or serious dispute before the early sixteenth century.

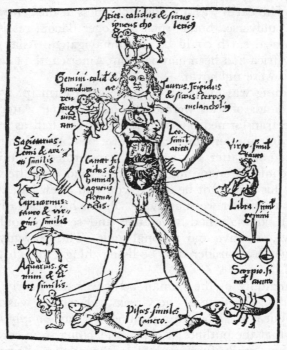

FIG. 19.—The correspondence between the parts of the body and the signs of the Zodiac. (*Margarita Philosophica*. Reisch. 1508.)

Examples of the Science of the Middle Ages

MEDIÆVAL NATURAL HISTORY, GOOD AND BAD

Job 4, 11 contains the words "*lish* perished for lack of food". The word *lish* has been variously translated as old lion, tiger, panther, pard, in various versions of the Bible. The Septuagint Greek version translated it μυρμηκολέων which literally means *ant-lion*, and the old Latin version of the Bible used the same word, *myrmicaleon*.

A very popular work, *Physiologus*, the original of which dates perhaps from A.D. 300, is composed of moral stories about beasts: the author of it was delighted with the ant-lion, and, using his imagination, produced the following story.

CONCERNING THE ANT-LION

Physiologus said that from the ant and the lion is born a beast which is called the ant-lion. And this animal when born swiftly perishes, because it cannot avail itself of food, but is unable to take it and dies of hunger. And that this is true Holy Writ is the witness, saying "The ant-lion perished for lack of food". For being of two natures, whenever he wishes to take meat, the nature of the ant, which

FIG. 20.—Illustration of various beasts from Bartholomew's *De Proprietatibus Rerum*. The work was written about 1260: the picture is taken from an edition printed by Wynkyn de Worde in 1495.

has an appetite for seeds, refuses the meat; but when he wishes to be nourished on seeds, the nature of the lion resists him. So being unable to eat either meat or seeds he perishes. Such are those who wish to serve two masters, God and the devil, God teaching them to be pure and the devil persuading them to be licentious. . . .*

But Albertus Magnus had studied that most interesting ant-trapping insect, now called *Myrmeleon formicarius*, still known as the ant-lion (or in the U.S.A. the doodle-bug), whose habits you may find narrated in the *Encyclopædia Britannica*. Albertus tells us, quite correctly (*De Animalibus*. xxvi, 20):

The formicaleon is called the ant-lion, which is also called murmicaleon. To begin with this animal is not an ant, as some say. For I have had a great deal of experience of it and have shown my colleagues that this animal has very much the shape of a tick, and it hides itself in the sand, digging in it a hemispherical cup, at the bottom of which is the ant-lion's mouth: and when ants, bent on gain, cross the pit, it seizes and devours them. This we have very often watched. In winter also it is said† to rob the ants of their food, for it gathers nothing for itself in summer.

PETER PEREGRINE DESCRIBES THE MAN OF SCIENCE IN HIS TREATISE ON THE MAGNET

But know, dearest friend, that the investigator in this subject must understand nature and must not be ignorant of the celestial motions: and he must himself be very diligent in manual operations, to the end that through operations with this stone (the loadstone) he will be able in a short time to correct an error, which he could not possibly do by means of his knowledge of nature and mathematics, if he lacked careful-

---

* Translated from the Greek version. *Cod. Vind. Theol.* 128, *c.* 20. A different version is given in the English translation of W. Rose, *Epic of the Beast*. Routledge, 1924, a most amusing work.

† St. Albert is exceptional among mediæval writers in distinguishing what he has seen from what he has been told. This part of the story is not in fact true. Isidore of Seville (c. A.D. 600) knew of the habits of the ant-lion, but does not state that he himself has witnessed them.

ness in the use of his hands. For in operations whose mechanism cannot be seen, we search out much by manual industry and, for the most part, without it we can make nothing perfect or complete. Yet there are many things subject to the empire of reason which we cannot completely investigate by the hand. (*Epistola de Magnete*, Petrus Peregrinus. Modified from the translation of S. P. Thompson.)

ROGER BACON IN 1269 FORECASTS THE SCOPE OF SCIENCE

And when we wish, things far off can be seen as near, and *vice versa*, so that at an incredible distance we might see grains of sand and small letters and the lowest things may appear very high, and *vice versa** and hidden things be seen openly and open things be hidden, and one thing may be seen as many and *vice versa*, so that many suns and moons may be seen by the artifice of this kind of geometry. . . .

And the fifth part is of the making of instruments of marvellous excellence and utility, as instruments of flying and of moving in chariots without animals at an incomparable velocity, and of navigating without oars more swiftly than can be supposed possible by the hands of men. . . . Flying-machines can be made, and a man sitting in the middle revolving an engine by which skilfully made wings beat the air like a flying bird. (Sloane MS., fol. 83, b. 1 & 2.)

ROGER BACON IN 1248 ANNOUNCES THE COMPOSITION OF GUNPOWDER

In his *Secret Works of Art and Nature*, c. 1248, Bacon says: "Sed tamen salis petre LVRV VOPO VIR CAN VTRI et sulphuris et sic facies tonitruum et coruscationem si scias artificium." The words in capital letters have been read as an anagram of R. VII. PART V, NOV. CORUL, V or, in full, 'recipe vii partes, v novellæ coruli'. Thus the passage means "but take 7 parts of saltpetre, 5 of young hazelwood (as

---

* It does not seem improbable that Bacon in experimenting with lenses made a crude Keplerian telescope of two convex lenses: this would magnify and invert as he asserts.

charcoal) and 5 of sulphur, and so you will make thunder and lightning if you know the trick", which 'trick' was the purification of the saltpetre and diligent incorporation of the materials. The words LVRV etc., which are found in the printed editions, are curiously enough not contained in any of the known MSS. which have in place of them a different cypher, doubtless also a recipe for gunpowder.

In 1269 when he wrote his *Greater Work* the secret was generally known, but gunpowder was still a toy. He tells us:

"And we have experience of this from that boys' prank which is done in many parts of the world, namely that by a contrivance no bigger than a man's thumb and the violence of that salt which is called saltpetre, so horrible a noise is made by the bursting of such a little thing, a mere bit of parchment, that it seems to exceed the loudest thunder and in its brightness it exceeds the biggest flash of lightning." (*Opus Majus*, R. Bacon. Ed. J. Bridges. Vol. II, 218.)

He is evidently describing a cracker made of a tube of rolled-up parchment filled with powder. The Middle Ages were not used to explosions and somewhat exaggerated their effects.

RULES OF HEALTH OF THE TWELFTH CENTURY

Rule of Living

> *The Salerne Schoole* doth by these lines impart
> All health to *Englands King*,* and doth advise
> From care his head to keepe, from wrath his hart
> Drinke not much wine, sup light, and soone arise,
> When meat is gone, long sitting breedeth smart;
> And after noone still waking keepe your eies,
> When mov'd you find yourself to *Natures Needs*
> Forbeare them not, for that much danger breeds,
> Use three Physitians still, first Doctor *Quiet*,
> Next Doctor *Mery-man*, and Doctor *Dyet*.

Remedies against Poison

> Six things that heere in order shall insue
> Against all poysons have a secret poure

* The king in question is not identified.

Peares, Garlicke, Reddish-roots, Nuts, Rape, and Rew
But *Garlicke* cheefe, for they that it devoure,
May drink, and care not who their drink do brew,
May walke in ayres infected* every houre:
Sith Garlicke then hath poure to save from death:
Beare with it though it make unsavoury breath:
And scorne not Garlicke like to some, that think
It onely makes men winke, and drinke, and stink.

It is to be noted that the rules of health are admirable, but
the drugs useless.

(From the *Regimen Sanitatis* of the School of Salerno. The
Latin original dates from the twelfth century, and the English
translation was made in 1607 by Sir John Harington, inventor
of the water-closet.)

ANGLO-SAXON REMEDIES FROM THE LEECH-BOOKS
(eleventh century or earlier

Against lice: pound in ale oak-rind and a little worm-wood,
give to the lousy one to drink. Against lice: quicksilver and old
butter; one penny weight of quicksilver and two of butter;
mingle all together in a brazen vessel.

In case a man be lunatic; take skin of a mere-swine or porpoise, work it into a whip, swinge the man therewith, soon he
will be well. Amen.

If thou be not able to stanch a blood-letting incision, take
new horses' dung, dry it in the sun, rub it to dust thoroughly
well, lay the dust very thick on a linen cloth; wrap up the
wound with that.†.

(From *Leechdoms, Wortconning,* and *Starcraft* of Early England.
Rolls Series. Collected and edited by the Rev. Oswald Cock-
ayne. Longman, Green, Longman Roberts and Green.
London, 1865.)

---

* i.e. by foul vapours or evil celestial influences.

† This remedy is calculated to produce tetanus in a large proportion of cases.

ALCHEMY AT ITS BEST

Section I. *On the materials used in the Art of Chemistry*

Know, may Allah have mercy on thee, that the materials used in the Art of Chemistry are of one species essentially. They are called the metallic minerals and subdivided into six sorts varying in form and in properties, but not immutable as are individual animals and plants.

They are gold, silver, copper, iron, lead and tin. Each of them is marked off from the others by accidental distinguishing properties, and it should be possible to effect the necessary removal of these properties, the specific nature remaining constant. We say and maintain that two species of natural things which differ radically and essentially cannot be changed and converted one into the other by the Art, as for example, man and the horse. But these six bodies can be mutually converted: thus lead may be converted into silver, for if you place a pound of lead in the fire, it rectifies and matures it, and most of it is burnt away, leaving a small part as silver—about a quarter of a drachm of pure silver from every pound of lead.*

Now since it is possible for a part of the lead to be changed into silver, there is nothing to hinder the conversion of the whole. In the same way silver may be converted into gold, by the refinement of the smelting fire only. For it is tinctured by the fire and strengthened and transmuted and behaves like gold with the touchstone. Thus it is possible to effect a certain transmutation since the specific nature is constant; but if silver differed from gold in species it would not be possible to convert it into it, just as it is impossible to convert a horse into the human species by Art, because they differ radically and essentially.

(From the *Book of Knowledge Acquired Concerning the Cultivation of Gold*, by Abu 'l-qasim Muhammad Ibn Ahmad Al-'iraq. Translated by E. J. Holmyard. 1923. Librarie Orientaliste, Paul Geuthner, Paris.)

---

* This is a correct observation. All lead as smelted contains a small proportion of silver.

PLATE V

Initial letter, from Vesalius, showing the experiment
described on page 67–8.

PLATE VIb

Title-page showing sailors using the cross-staff (top right), cross-bow (bottom left) and quadrant (bottom right).

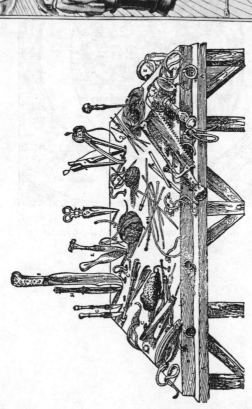

PLATE VIa

The instruments used by Vesalius for dissection (1543).

A BRIEF ALCHEMICAL TEXT OF THE SYMBOLIC TYPE

The vision of Sir George Ripley; Chanon of Bridlington.

When busie at my booke I was upon a certeine night
This Vision here exprest appear'd unto my dimmed sight,
A *Toade* full rudde I saw did drinke the juce of grapes so fast
Till over charged with the broth, his bowells all to brast;
And after that from poysoned bulke he cast his venome fell,
For greif and paine whereof his Members all began to swell,
With drops of poysoned sweate approaching thus his secret Den,
His cave with blasts of fumous ayre he all be-whyted then;
And from the which in space a golden humour did ensue,
Whose falling drops from high did staine the soile with ruddy
    hew;
And when this Corps the force of vitall breath began to lacke,
This dying Toade became forthwith like Coale for colour blacke:
Thus drowned in his proper veynes of poysoned flood,
For tearme of eightie dayes and fowre he rotting stood,
By tryall then this venome to expell I did desire,
For which I did committ his carkase to a gentle fire:
Which done, a wonder to the sight, but more to be rehear'st,
The *Toade* with Colours rare through every side was pear'st.
And White appeared when all the sundry hewes were past,
Which after being tincted Rudde, forevermore did last.
Then of the venome handled thus a medicine I did make;
Which venome kills and saveth such as venome chance to take.
Glory be to him the graunter of such secret wayes.
Dominion, and Honour, both with Worship, and with Prayse.
                                                    AMEN.

(From *Theatrum Chemicum Britannicum*, Elias Ashmole, 1652.)

THE BRITISH STANDARD YARD

(The date of this statute is not accurately known but is
before 1284.)

"Three grains of barley, dry and round, make an inch; twelve
inches make a foot; three feet make an ulna (yard); five & a

half yards make a perch; and forty perches in length and four perches in breadth make an acre.

And it is to be remembered that the Iron Ulna of our Lord the King contains three feet and no more; and the foot must contain twelve inches, measured by the correct measurement of this kind of ulna, that is to say, one thirty-sixth part of the said ulna makes one inch neither more nor less. . . ."

(It is seen that the natural standard of the barley-corns was already replaced by an artificial standard bar.)

(Quoted from *British Weights and Measures*, by Sir C. M. Watson. Murray, 1910.)

CHAPTER FIVE

The Revival of Observation

Science revives

Science has had its periods of advance and regress, yet since the middle of the fifteenth century it has been in continuous advance. There have been, it is true, some false trails and blind alleys, but work has never slackened nor interest declined. The Middle Ages were not so barren of scientific achievement as is generally thought, yet it is obvious that between 1450 and 1500, there was a rapid intensification of scientific interest. The scholastic philosophy had shown the men of the Middle Ages the world under the guise of purpose, but had done very little to describe it in terms of number, weight and measure. The desire for more knowledge of the world came from three chief sources: first, a new study of the learning of Greece; secondly, a growing dissatisfaction with the inaccuracy and small extent of man's knowledge of the world; lastly, from the growing interest which attached to industry, as its products became socially more important.

In the Middle Ages a knowledge of Greek was a rare accomplishment, and so the Greek authors, by far the greatest men of science there yet had been, were known only partially and in Latin translations, often made from Arabic versions. The new study of Greek began in Italy and became of importance from about 1450 onward: at the same period was developed the art of printing, so that in the fifteenth century there began both the new study and the wide diffusion of Greek scientific authors. Aristotle's biological works, Pliny, Euclid (in Latin), some of Ptolemy, Theophrastus, Dioscorides and a number of medical authors appeared before 1500: Galen and Hippocrates, Euclid in Greek, and Ptolemy's Almagest round about the 1520's; and a number of mathematical works, including those of Archimedes, about the middle of the sixteenth century.

The readers of these texts discovered at once that the ancient authors disagreed with or went beyond the traditional mediæval sources, a fact which shook their authority: and later, that the facts of Nature, when observed, disagreed with both. There was growing up in Europe, moreover, a large, wealthy, and influential merchant class, which drew much of its wealth from manufactures, and was therefore interested in the scientific study of industrial processes—salt-works, mining, metallurgy, pharmacy, perfumery, pyrotechnics and the like. The ancients, the Arabs, and the scholastics, alike were silent on these processes. Moreover, the world was expanding in extent and wealth through exploration and trade, and facts were coming to hand from foreign parts concerning which the old authorities could tell nothing. It was of no use to consult Aristotle about the new beasts of the Indies or Ptolemy on the geography of the Americas. There had come into being new subjects of knowledge which forced men to use their own eyes.

Astronomy. The Copernican theory

The first dawn of new observation came from the German astronomers Peurbach and Müller (Regiomontanus). The usual astronomical theory of the Middle Ages was that of Aristotle's concentric spheres; and Peurbach's work was to restore the much superior system of Ptolemy (pp. 29–30). He retranslated and reconstructed Ptolemy's Almagest from the only version available, the inaccurate Latin translation of the Arabic translation of the Greek original, and he started making fresh observations of the positions of the principal stars and mapping them on celestial globes. This gave him exact fixed points with reference to which to plot the positions of the planets, and he was thus able to construct better tables of their motions. Müller worked under Peurbach, and in 1460 went with him to Italy to learn Greek, which was being taught by the Byzantines, so that he might be able to retranslate Ptolemy from the Greek. When he came back to Nuremberg he set up a platform-observatory on the roof of the house of his friend and patron Bernard Walther: this observatory has been thought to be the scene of the *Melancholia*, a famous engraving by Dürer, who later occupied the house. He, and Walther after him, continued to observe till 1504; thus their great successor, Nicolaus Copernicus, who began observing about 1507, could draw on sixty years of modern observations as well as those of himself and of Ptolemy.

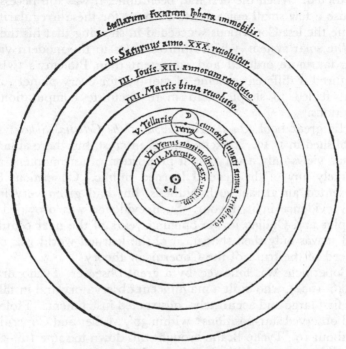

FIG. 21.—The system of the world according to Copernicus (1543).

Peurbach and Müller were content with an adaptation of Ptolemy's system of eccentrics and epicycles, but Copernicus made it his life's work to put forward and support the view, which he had derived from certain of the Greeks, that

(1) the sun and stars were motionless at the centre and circumference of the universe respectively,

(2) that the earth rotated on its axis in 24 hours,

(3) that the earth and the planets revolved about the sun, while the moon revolved about the earth.

With the laborious methods of calculation then in use, it was a most difficult task to harmonise the earlier (and often erroneous) observations with this scheme. The system was not a perfect one. It was not possible to assign any uniform motion to the planets which would enable their position to be exactly predicted. When the best had been done, it was still necessary to use a few small epicycles to account for these irregularities. None the less Copernicus succeeded in showing that his theory of the solar system, which approximates to the modern view, was far more ordered and consistent than Ptolemy's (which required a different scheme of motions for every planet), and that it led to simpler and more accurate computation of positions.

His great book *On the Revolution of the Celestial Spheres* was published in 1543. It excited much interest, but there adhered to his views only a very small proportion of astronomers, and scarcely any of the general learned public. Copernicus had presented an attractive hypothesis, but had given very little new evidence in its favour. Such evidence was provided by Kepler and Galileo in the opening years of the next century; and it was only then that most of the learned world was convinced of the truth of the Copernican theory.

Copernicus was followed by a great observer, Tycho Brahe (1546–1601), who built a magnificent observatory and installed the first large and accurately-constructed instruments. Ptolemy had observed star-positions* within 30', Müller and Copernicus to about 10', Tycho Brahe brought this down to some 10"–20". He did not agree with the views of Copernicus but suggested that the earth was stationary, that the sun revolved about it, while the other planets revolved about the sun. His system found little favour, but his observations provided a basis of accurate measurements which enabled Kepler to discover the simple and beautiful laws that express the planetary motions (p. 115). In these astronomers of the fifteenth and sixteenth centuries we see three of the characters of true scientists:

(1) they *observe for themselves,* seeking ever higher accuracy:

* 30' is roughly the arc subtended by the moon.

(2) they *abandon preconceptions* about the nature of the universe:

(3) they keep *accurate records* of their work and publish them to all.

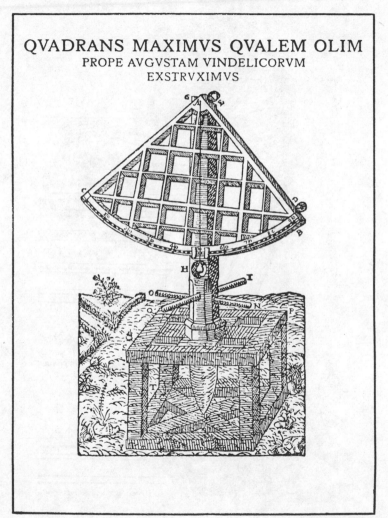

FIG. 22.—A huge quadrant erected by Tycho Brahe.

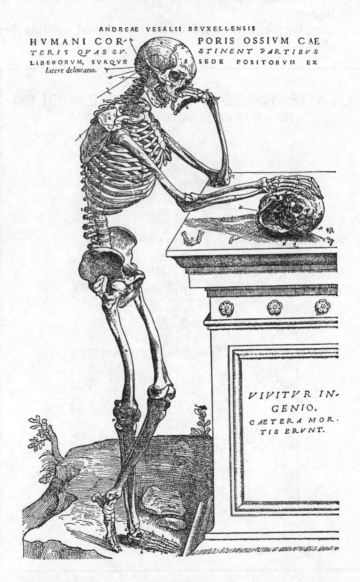

FIG. 23.—A page from Vesalius' *De Fabrica Humani Corporis.*
Note that the skeleton is depicted in its living posture.

The Biological Sciences—Andreas Vesalius

The only sciences which in this period made progress at all comparable with astronomy are the biological, especially human anatomy. Rather perfunctory anatomical demonstrations had continued throughout the Middle Ages, but in the early sixteenth century a few men began to observe exactly

FIG. 24.—The artists making the drawings for the herbal of Fuchs, published in 1542.

and draw accurately. Several artists of the XV and XVI centuries made dissections in order to study artistic anatomy. Leonardo da Vinci, artist and scientist, dissected some thirty bodies and left note-books containing wonderful anatomical drawings; but as these were not published they had no general effect. The influential anatomist was Andreas Vesalius, who in 1543 published his chief work. "De Fabrica Humani Corporis", *On the Fabric of the Human Body.* This is a completely new departure. The tone is that of modern science: change but a few terms and long passages could appear without strangeness in a modern scientific journal. He is not concerned to *prove* any general theories, simply to find out and record the structure of the body. He even carries out physiological experiments, one of which is translated on pp. 84–85. The illustrations are superb (Fig. 23, Pls. V, VIa), and indeed

the noble wood-cuts and engravings of the sixteenth century
were a great contribution to scientific record. Vesalius and
other great anatomists of the time cleared up and recorded
the main points of gross anatomy that could be seen with the
naked eye; for there were as yet no microscopes, by which
the fine structures could be elucidated.

In the other biological sciences, a similar work was done,
that of exact observation and beautiful pictorial record, but it
cannot be said that any great progress was made in physiology.
—the understanding of the working of animals and plants:
botany and zoology books were no more than herbals and
descriptions of beasts, although as the century went on these
became clearer and less credulous. In medicine, the physiology
of Galen (p. 34) still reigned supreme, and the genuine advances

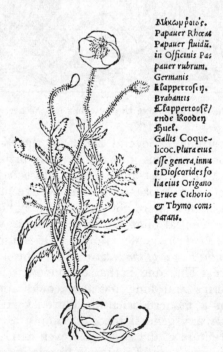

Μήκων ροιάς.
Papauer Rhœas
Papauer fluidū.
in Officinis Pa
pauer rubrum.
Germanis
klapperrofen.
Brabantis
Clapperroofe/
ende Rooden
Huel.
Gallis Coque-
licoc. Plura eius
esse genera, innu
it Dioscorides fo
lia eius Origano
Eruce Cichorio
& Thymo cons
parans.

FIG. 25.—Drawing of a poppy from a
sixteenth century herbal.

in treatment were not great. Paracelsus stirred up the medical faculty by urging in violent language the use of mineral medicines: but in fact most of the medicines he recommended were inefficacious and he did little more than awake the faculty of medicine to the possibility of departing from the practice of such authorities as Galen, Hippocrates, Avicenna. Ambroise Paré, however, went far to improve surgery: amongst other things may be mentioned his use of the ligature to stop bleeding in operations and his abolition of the treatment of wounds with boiling oil.

Chemistry and Physics in the sixteenth century

The most backward of the sciences were chemistry and physics. Chemistry had not yet become a science. There was *Alchemy*, of much the same type as that in earlier centuries: there was *Metallurgy*, on which Agricola (Georg Bauer) and Lazarus Ercken wrote magnificent treatises: there was *Pharmacy* on which Paracelsus and many others produced bulky works; there were books on technical subjects, such as fireworks, explosives, distillation of perfumes and spirits, the making of glass, porcelain, etc.: but none of these were more than accounts of crafts which had to do with materials, and as yet there was no Chemistry—no single Science of the Preparation and Properties of Materials. The first book which can be called a text-book of Chemistry because it combined all these crafts and united them in a single scheme, is the *Alchymia* of Andreas Libavius, published in 1597. There was, however, not a little progress; for the sixteenth century, with its wealth and manufactures, required a large variety of materials and so, although there was as yet no science of chemistry, a great deal of chemical technique accumulated for the chemists of the next century to discuss.

The physics of the sixteenth century made little advance on that of the Arabs. It amounted to no more than a few minor works on the kind of physics which the Greeks had studied, namely, optics, machines, and mechanical powers, floating bodies and centres of gravity; and it was not until the last few years of the century, that two great men, William Gilbert.

the "father of magnetism", and Simon Stevin, a Dutch mathematician and physicist, produced books that began the new
period of experimental physics. These will be treated in
Chapter V with the early work of the seventeenth century, of
which they really form a part.

FIG. 26.—The beginning of analytical chemistry. The testing of
nitre at a sixteenth century works.

Mathematics before 1600

Mathematics made great advances in these years, so much
so that mathematical methods were ready to hand as soon as
science was ready to use them. First of all we must remember
that the various translations of Euclid gave the world much
of the Greek mathematics in a clear form. Algebra had been
introduced by the Arabs, but in 1500 most of what we would
do by algebra was done by geometrical methods. The successive powers of numbers x, x^2, x^3, were thought of as lines, areas,
and solids; and so the notion of x^4 was as difficult to them as
the fourth dimension to us, because they could not imagine it

Fig. 27.—The use of counters and a counting board for arithmetic in the sixteenth century. A shop's counter is so called because it was formerly marked for the use of counters.

geometrically. Yet by the end of the sixteenth century mathematicians could solve not only quadratics, but also cubics and biquadratics; they had something like the modern algebraic notation: they had a very good knowledge of Euclidean geometry, conic sections and higher curves; and a fair acquaintance with trigonometry.

In the fifteenth century ordinary people worked out sums with an abacus or counters, or by clumsy methods using Roman figures; and *algorism*, as working in Arabic figures was called, was still a learned accomplishment. In the sixteenth century educated people learnt to use Arabic figures and the multiplication table, though a good many knew it only to five-times. The labour of astronomical calculations was enormous, and only at the beginning of the seventeenth century was it eased by the invention of logarithms.

Navigation

Lastly, we must not omit new progress in navigation. Before 1450 there was little but coastal sailing, but the ocean voyages to America and beyond raised new problems. Sailors had to learn about the variations of the compass: they had to learn to take the height of the sun with crude instruments such as the cross-bow or the cross-staff (Pl. VIb) and so to find their latitude. The longitude could not be found except by astronomical observations on shore, so ships had to use dead-reckoning, the inaccuracies of which are made clear enough in the extracts cited on pp. 86–7. Indeed no ship could find its longitude by observation until about 1760, when the first accurate chronometers were made. The early charts made no allowance for the spherical form of the earth. These did well enough for short distances in low latitudes; but as oceanic navigation became frequent something better was needed, and Mercator's great map, published in 1568, was the first to try to allow for this by increasing the distance between the parallels in high latitudes; yet this method was not made mathematically exact until Edward Wright worked out the theory in 1590.

Science in the sixteenth century

Summing up the century and a half from 1450 to 1600, we see practical improvements and expansion of knowledge in almost every science. But there is little attempt to assess the value of evidence or to put knowledge into system and order, and theory—except in astronomy—lags vastly behind fact. The ancients were no longer considered as reliable authorities, yet nobody laid down what was to be believed as scientific fact and what was not; nor did anyone propose any *method* of going about scientific problems. Credulity remained gigantic. All sorts of magical sympathies and antipathies still ranked among natural laws: the most unlikely travellers' tales and traditional lore still gained credence; and it was common enough to find men who practised or at least treated of magic and science together: Paracelsus, Dr. John Dee and Baptista della Porta afford us examples. Science needed to sort out its valuables and set them in order—and this was one of the many achievements of the seventeenth century.

Examples of Sixteenth-Century Science

COPERNICUS EXPOUNDS THE MOTIVES THAT LED
TO HIS WORK

... So I would not have it unknown to your Holiness that what moved me to consider another way of reckoning the motions of the heavenly spheres was nothing else than my realisation that Mathematicians were not agreed about the matter of their research. For, first of all, they were so uncertain about the motion of the Sun and Moon that they could not demonstrate or observe the unvarying length of the yearly cycle. Then, in setting out the motions of the Sun and Moon and of the five other planets they did not in each case employ the same principles, assumptions and demonstrations of apparent revolutions and motions. For some used concentric circles only, others eccentrics and epicycles, by which, none the less, they were not able to fully arrive at what they sought. For those who believed in concentric circles,* although they demonstrated that several different motions could be compounded from them, could not derive from these motions anything certain that really corresponded to the phenomena. Those who devised eccentric circles,† though they seemed to have accounted for the visible motions in great part when their figures agreed with them; yet meanwhile they admitted that the most part seemed to contravene the first of principles, that of the equality of motion. They could not find therein or deduce another most important thing, that is the form of the world and the certain symmetry of its parts.‡

... And so after long consideration of the uncertainty of the mathematical traditions about the inferring of the motions of the spheres, it began to vex me that no more certain system of the motions of the machine of the world, which was created for us by the most good and orderly Artificer of all things, could be agreed on by the philosophers. ...

* Eudoxus, Callippus, Aristotle.

† Hipparchus, Ptolemy.

‡ Copernicus takes it as self-evident that the motions of the heavenly bodies are uniform and symmetrical.

(He then tells us he re-read the authorities and found that some of the ancients supposed the earth to move, and continues . . .)

. . . And so, taking occasion from this, I too began to think about the mobility of the earth. And although it seemed an absurd opinion, yet, because I knew that others before me had been granted the liberty of supposing whatever orbits they chose in order to demonstrate the phenomena of the stars, I considered that I too might well be allowed to try whether sounder demonstrations of the revolutions of the heavenly orbs, might be discovered by supposing some motion of the earth. And so, by supposing those motions* of the earth, which I set out in the work that follows, I discovered by much and long observation, that if the motions of the rest of the planets are compared with the motion of the earth, and are computed as for a single revolution of each planet, not only do their appearances† follow therefrom, but the system moreover so connects the orders and sizes both of the planets and of their orbits, and indeed the whole heaven, that in no part of it can anything be moved without bringing to confusion the rest of the parts and the whole universe.

(From the preface of *De Revolutionibus Orbium Coelestium*, Nicolaus Copernicus, 1543, addressed to Pope Eugene IV.)

ANDREAS VESALIUS DEMONSTRATES THE CONNECTION OF THE VOICE WITH THE RECURRENT NERVES

[This is done by dissection of a living sow. This animal is chosen because it can be counted on to squeal continuously and so provide the voice which is the subject of the experiment.]

Before the animal is thus tied down, I usually address myself to any spectators who are not yet versed in the anatomy of dead animals and go over the things that should be seen in the dissection in question, in order that a wordy explanation in the middle of the dissection shall not interrupt the work, and so

* Daily rotation, and yearly revolution about the sun.

† i.e. "the actual positions as observed can be deduced from the motions supposed."

that it shall not come to be disturbed by talking. And so I immediately make a long incision in the neck with a sharp razor, so as to divide the skin and the muscles beneath it as far as the windpipe; taking care that the incision does not go too far to the side and injure an important vein. Then I take hold of the windpipe in my hands and stripping it from the attached muscles with my fingers only, I look for the carotid arteries at the side of it, and for the nerves of the sixth pair of cerebral nerves stretched out thereon. Then I observe the recurrent nerves also attached to the sides of the windpipe, and these I sometimes compress by ligatures and sometimes sever: this I do first on one side so that when the nerve is compressed or severed it may be clearly observed that half the voice is lost, and that when both nerves are damaged the voice is completely lost, and that if I loosen the ligatures it returns.

(Note the clear description and the experimental character of the work. See illustration in Plate V.)

(From *De Corporis Humani Fabrica*, Vesalius, 1542, p. 668.)

MEDICAL PRESCRIPTIONS IN 1557

An oil for closing wounds

In a pound of olive oil cook ten green lizards and filter through linen, adding thereto one measure of marjoram and wormwood, which gently cook and set by for use.

Medicament for recent wounds

Take earthworms* washed in wine and place them in a closed jar. Cook them in a double vessel for a day, and when they are liquefied add either properly prepared balsam, or resin of the fir or larch tree. It heals any new wound in a short time and especially wounds of the head.

Plaster to mature, extract, and clean abscesses

Take a white onion, cooked in the ashes, and an equal amount of old hog's lard; rub them together and add wheat flour. Some people also add calves' gall.

* Oil of Earthworms remained a standard remedy for wounds until the eighteenth century.

For tooth-ache, taking away the pain very quickly

Parsley seed, 1 scruple, opium half a scruple, henbane* two scruples; make pills, of which place one in the tooth.

An Electuary to stop bleeding, strengthen the heart, purge yellow bile, phlegm and some black bile, without inconvenience.

Candied roses†	2 oz.	Picked rhubarb root	2 drachms
Candied Borage	1 oz.	Picked Manna	1 oz.
Candied Violets	½ oz.	Jalap ⎫ of each Senna ⎭	½ oz.

Simple Electuary
  of Prunes
	2 oz.	Selected ground
		  cinnamon
			½ drachm

It is made into a syrup with lemon juice: a large spoonful to be taken fasting.

(A seventeenth-century hand has written *Electuarium Nobile* beside this in the copy used by me. It is in fact a strong purgative, in all ages the stand-by of physicians.)

(*De Compositione Medicamentorum*, Fumanelli. 1557.)

EDWARD WRIGHT, IN 1610, BEWAILS THE DIFFICULTIES
OF THE MARINERS OF HIS TIME

Nautical Observations

There were . . . Old Shipmasters, who not many yeares since had mocked them that used charts or cross-staves, saying they cared not for their sheepskinnes, they could keep a better account upon a boord: and them that observed the sun or starres for finding the latitude they would call sun-shooters‡ and star-shooters, and ask if they had hit it. But mark what commeth hereof, for one of these Masters was he, as I take it, of whom an ancient seaman (yet living as I think) once told me, who having undertaken the charge of conducting a ship

---

* Henbane contains hyoscyamine, a powerful local anæsthetic and narcotic. This would presumably be effective, though possibly dangerous.

† Sugar of Roses. A material made by boiling down rose-petals and sugar or honey.

‡ Note the attitude of the observer with the cross-bow in Plate VIb. The phrase 'shooting the sun' is still used by sailors for taking observations with the sextant.

from England to St. Michael's (the Easternmost of the Azores) and after long seeking not able to find it, for shame and sorrow cast himself overboard.

Elizabethan Charts

Hereto accord the often experiments and usuall practise of many well experienced and judiciall mariners and seamen of our time, who confesse, that in sailing from the West Indies to the Azores, they have often fallen with those Ilands, when by their account* according to the Chart they should have beene 150 or 200 leagues to the Westwards of them. . . . For who that loveth truth, can patiently endure to heare the Mariners common and constant complaint of 150 or 200 leagues error in the distance between the Bay of Mexico and the Azores, or (that which is yet most intollerable and monstrous) of 600 leagues difference in the distance between Cape Mendosino† and Cape California, some making that distance to be twelve or thirteen hundred leagues, where other, and that more probablie, make it to be no more than six or seven hundred.

(*Certaine Errors in Navigation*, Edward Wright, 1599.)

SEVEN TIMES EIGHT

. . . Because that Great Summes cannot be multiplied but by the multiplication of dygetes; therefore I thinke best to shewe you fyrste the art of multiplyenge them as when I say 8 times 8 or 8 times 9. And as for the small dygetes under 5 it were but foly to teach any rule, seynge they are so easy that every chylde can do it: but for the multiplycation of the greater dygetes thus shall you do. Fyrst set your dygetes out over the other ryght, then loke how many eche of them lacketh of 10 and wryte that agaynst eche of them, and that is called the dyfferences, as yf I wolde knowe how many are 7 tymes 8, I must write those digettes thus.

8

7

* Reckoning of distance travelled and direction.
† The distance between Cape Mendocino (lat. 40 N. 124 W. appr.) and Cape California is about 1300 geographical miles, which is about 430 leagues of 3 miles.

Then do I loke how moche 8 doth differ from 10, and I fynde it to be 2, that 2 do I write at the ryghte hand of 8, thus 8   2. Then do I take ye dyfference of 7, lykewayse from 10, which is 3, and I write that at the right side of 7, as you se in this example:

$$\begin{array}{cc} 8 & 2 \\ \underline{7} & \underline{3} \end{array}$$

Then do I drawe a line under them, as in addition, thus. Then do I multiply the two differences, sayeng: 2 tymes 3 make 6, that I must ever set under the differences beneath the line; then must I take the one of the differences (which I wyl for all is lyke) from the other digette, not from his owne and that which is lefte, must I write under the digettes, as in this example:

$$\begin{array}{cc} 8 & 2 \\ \underline{7} & \underline{3} \\ 5 & 6 \end{array}$$

(The reader may now figure out why it works—a very simple problem in algebra.)

(From *The ground of artes teachyng the worke and practise of Arithmetike*, by Robert Recorde, 1542.)

SUPERSTITION IN SCIENCE

But if anyone shall objecte that Words and Characters have no virtue; and say as well as others, That they are of no more power than a bare Mark or naked Cross or Signe: . . . let him tell me, who believeth such things, whence it comes to pass, That Serpents in Helvetia, or Suevia, do understand those Greek words, Osii, Osiia, Osii; since the Greek tongue is not so vulgar in those countries, that venemous worms should understand it, or in time learn it? How should they come to understand them, or in what Universitie have they learned them, that as soon as ever they hear these words, they will immediately stop their eares with their tayles, that they may not hear them again?

(From a seventeenth-century translation of a work, *Of Occult Philosophy*, attributed to Paracelsus and dating from the mid-sixteenth century.)

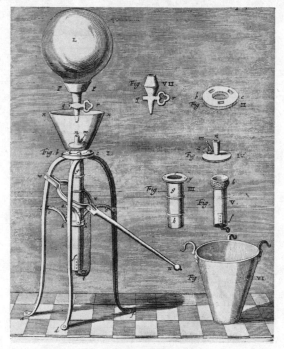

PLATE VII

One of von Guericke's later air-pumps. In order to
keep air from leaking in, it worked in a conical
bucket of water (marked Fig. VI) and the joint of
the globe and tap was also water-sealed.

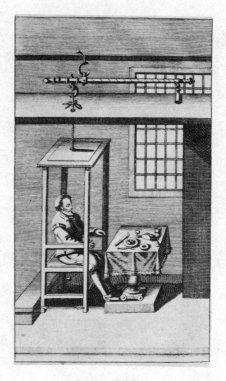

PLATE VIII

Sanctorius in his chair, suspended
from a large steel-yard. He thus
showed that the loss of weight by
perspiration was greater than that by
other excretions.

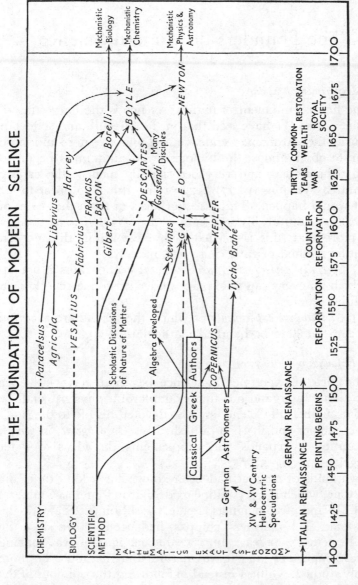

THE FOUNDATION OF MODERN SCIENCE

Fig. 28.

CHAPTER SIX

The Founders of Modern Science

The Experimental Method

The first requirement of natural science is the harvesting of reliable facts. We have seen that in the fifteenth and sixteenth centuries the astronomers endeavoured to make more and more accurate observations; the biologists and anatomists began to see for themselves, and very occasionally men of the calibre of Andreas Vesalius (p. 77) began to do things in order to see what would happen. This last step takes us from the realm of *observation*—natural history—into that of *experimental science:* but more than this is needed before we attain the modern scientific method of gathering facts, namely:

(1) *Specially devised experiments* whereby apparatus is set up in which an event can take place under certain special known conditions, and

(2) *Quantitative experiments* of which the results are *numerical* and can therefore be handled mathematically.

The Refounding of Physics

The idea of experimental science began to have influence after about 1590, and our first example of a physicist shall be William Gilbert of Colchester, who devoted his life to the study of the magnet and in 1600 published his *De Magnete*, in which he continues, expands and supplements the ideas of Peter Peregrine (pp. 64–5).

Two kinds of magnet had been known for a long time, the lodestone, which is a crystalline oxide of iron found as a mineral (Fe_3O_4), and steel magnets, made by rubbing steel with the lodestone. The mariners' compass had been known since the twelfth century at least, but oceanic navigation was making it most important, and it was natural that it should be more deeply studied. Gilbert instead of following the opinions of the ancients (which he quotes only to rebut) describes experiments that he had himself done. His most significant conclusion was that the earth itself is a magnet with poles near, but not coincident with, its geographical poles; and that the compass-needle was attracted to these, and not, as previously supposed, to some northerly star or to a lodestone mountain in the Arctic regions. He was not content to assert this, but used the truly scientific method of turning up a sphere of lodestone and showing that tiny compasses placed thereon behaved as compasses do at the corresponding parts of the terrestrial globe (p. 98). It

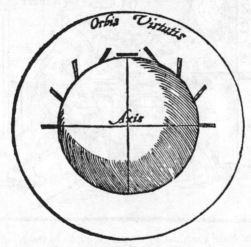

FIG. 29.—Gilbert's lodestone sphere or terrella, with iron wires to show dip. (*De Magnete*, 1600.)

is impossible here to recount the facts that he discovered, but their importance for the development of science was that they were based on deliberately contrived experiments, that they were published, and that they afforded examples of physical experiment to such men as Galileo and Bacon.

Galileo Galilei (1564–1643) was the first to employ the modern scientific method in its fullness. His work was in two fields, physics and astronomy; the second of these will be reserved to the following chapter. Galileo was an enthusiastic and versatile man, skilful with his hands, a fine writer and an able mathematician. His main achievement in physics was the founding of the science of mechanics. Aristotle's ideas on motion (p. 32) had been occasionally questioned, but still reigned supreme. A 2 lb. weight, other things being equal, was believed to fall twice as fast as a 1 lb. weight. Bodies of a given weight

FIG. 30.—Making a magnet by hammering iron in the N—S meridian.
(From Gilbert's *De Magnete*, 1600.)

falling through a given medium were supposed to have a fixed and more or less constant speed which was supposed to be inversely proportional to the resistance of that medium; but how that resistance was to be defined did not appear. Nobody had tried to measure the velocities of such bodies, let alone the forces that impelled them. Such terms as motion, force, movement were used almost interchangeably.

In setting up a science of mechanics, Galileo had first to confute the Aristotelean views. In the year 1586 Simon Stevin had published (in Dutch) an experiment (p. 100) which confuted Aristotle's notion that the speed of falling bodies was proportional to their weight, but his work was little known. Galileo did many similar experiments, though he does not record the exact circumstances:* but the great advance he made was to put forward a theory that bodies fell with an accelerated motion which could be expressed by a mathematical formula. There

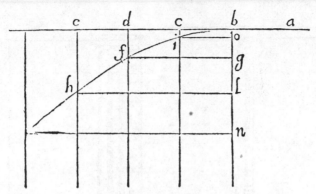

FIG. 31.—Galileo shows that a projectile discharged horizontally from *b*, will in equal time intervals describe equal distances horizontally, *bc*, *cd*, *de*; but will be accelerated vertically and will in each time-interval describe a greater distance, as *bo*, *og*, *gl*, *ln*. He shows that the path of the projectile *b i f h* must be a parabola.

were as yet no devices suitable for timing falling bodies, so he tried out his theory on bodies rolling on an inclined plane (p. 100): this is one of the earliest examples of a quantitative experiment, carefully checked and controlled. Moreover he saw that a smooth ball rolling on a smooth rising slope must slow down, and on a descending slope must speed up: this led him to the very important conclusion that a body in uniform motion on a flat level surface without friction, would so continue until some agency caused it to stop, whereas physicists had formerly considered with Aristotle, that some agency must continually act on a body in order to keep it in motion. Later in his life he combined these conclusions; he saw that a projectile, e.g. a cannon-ball fired over a cliff must continue to move horizontally at a uniform pace, and must also fall with an acceleration. He

* The story of his experiment of dropping weights from the leaning tower of Pisa may be true, but rests on slender evidence: there is no doubt, however, that he frequently dropped different weights from various eminences and observed that they struck the ground almost simultaneously.

then calculated mathematically that the path of a body which moved with these simultaneous motions was a parabola. Fig. 31 shows the diagram in which he expresses this conclusion. Here is a man of science experimenting, deducing laws, and building on these again.

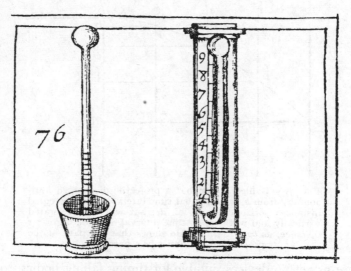

FIG. 32.—A picture, dating from 1626, of two types of thermometer The left-hand type was invented by Galileo, the right-hand probably by Cornelius Drebbel. (From *Annals of Science*, Vol. V, No. 2.)

This work on projectiles was but a small part of Galileo's physical discoveries. He showed how in machines (e.g. levers, screws, pulleys, etc.) decrease of working-distance was a necessary consequence of increase of force, and so dealt the first blow to the power-for-nothing dream which had always intrigued amateur engineers.

He studied a wide range of physical subjects and made important discoveries in all of them. He showed what was the relation between the density of a body and its power to float in a liquid: he was the first to show that the time of swing of a pendulum is (nearly) independent of the arc it swings through; he invented the thermometer; he was the first to assign a rate of vibration to a sound of known pitch; he attempted to measure the velocity of light and showed that it must be very great. His discovery of the telescope will be noticed in Chapter VII. Here we must mention him as originating the study of gases. He was the first to weigh air (p. 101) and probably the first to produce a vacuum, though he could not study its properties. Further material concerning Galileo is to be found in the following chapter.

This work led after his death to Torricelli's famous experiment, the making of the first mercury-barometer, which was the first clear proof that a vacuum could exist, and a confutation of Aristotle, who had proved from his incorrect assumptions concerning motion that it was impossible (p. 32). From Torricelli's time it was not long until von Guericke, burgomaster of Magdeburg, made the first air-pumps (Pl. VII) and astonished the world with the "force of the vacuum", as it was called.

During these years, 1590–1650, it was in physics that the greatest advances were made, but at the same time there was some important progress in physiology. About 1610 Sanctorius, an Italian physician, used a clinical thermometer—of a sort; he used pendulums for timing pulses: and, very much in the modern spirit, showed by weighing himself that the body was undergoing a continual loss of weight by "insensible perspiration" (Pl. VIII). Here we see quantitative experimental science—weighing and measurement—applied to the human body; and after Descartes had put forward the theory that the human body was no more than a mechanism guided by the soul, seated in a corner of the brain, others, such as Borelli, tried to give mechanical scientific explanations of the action of the body. In some respects, such as the inter-action of muscles, bones and joints, Borelli did valuable work, but in most of his explanations he went very far beyond what his knowledge of the facts warranted.

The Circulation of the Blood

But we have passed over the wonder of the age, Harvey's discovery of the circulation of the blood, made about 1615 and published in 1628. The heart had been thought of as, so to

speak, the source of life, and the seat of the emotions: here
was a man who showed it to be a pump, and who solved the
age-old problem of the motions of the heart and the blood, and
their relation to the lungs. The standard view of antiquity (see
Fig. 10, pp. 34–36) was that the blood ebbed and flowed in the
veins; and that this ebb and flow was chiefly operated by the
right ventricle of the heart. The left side was thought to
receive a small supply of blood through the central partition
of the heart, the septum (which, as Vesalius saw, has no apertures in it for blood to pass through), and to elaborate this
blood into the "vital spirits". Harvey discovered the actual
course of the blood, partly by studying the living heart in cold-
blooded animals, partly by reasoning from his knowledge of
anatomy. He saw that the valves of the heart were such as to
allow the blood to pass only in one direction. He worked out
how much blood the heart must be passing at each beat, and
concluded that in half an hour the heart propelled in this one
direction far more blood than its owner contained. Moreover,
it was evident to him that all the blood in the arteries came
from the aorta and ultimately from the veins. As an animal
in which an artery was opened was soon drained of blood, it
was clear that new blood was not being continually made to
supply the place of that which passed from the veins to the
heart; so Harvey was forced to conclude that the blood circu-
lated (p. 102). Here was an absolutely fundamental fact of
physiology, totally unguessed by Galen and Hippocrates, and
therefore very potent to inspire confidence in the new science
and distrust of the old.

The Statement of the Scientific Method

Such practical men as Stevin, Gilbert, Sanctorius, Harvey,
and even Galileo, did not altogether realise what they were
doing when they set about investigating Nature in what seemed
to them a reasonable way. But Francis Bacon (1561–1626),
who unlike these had little aptitude for practical science, understood the enormous importance of the new way of investigating
nature, and set to work to give an account of it. His first work
of this kind was *The Advancement of Learning* (1603) but we may

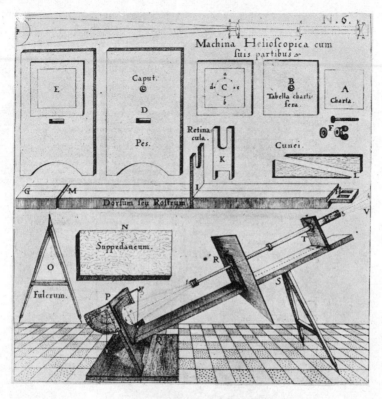

PLATE IX
Christopher Scheiner's apparatus for showing sun-spots by
projection (1630).

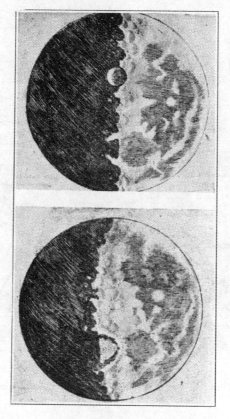

study it best in the *Novum Organum* (1620). This title means *The New Instrument*. Aristotle's logical works, the *Organon*, were the 'old instrument' and the scientific method was the new instrument: and just as such instruments as rulers and compasses aid the hand, so the scientific method was to aid the mind. Bacon starts by analysing the hindrances to gaining sound knowledge, namely the imperfections of our understanding, our habit of thinking about nature in human terms, our own prejudices and attachment to our favourite notions, the inexactness of language, and the impatience of men who want short cuts to complete explanations of the universe. He says that all these vicious habits must be "solemnly and for ever renounced, that the understanding may be thoroughly purged and cleansed: for the kingdom of man, which is founded in the sciences, can scarce be entered otherwise than the kingdom of God; that is, in the condition of little children".

So Bacon rejects all the knowledge of the ancients. Greek philosophy is "the talk of idle old men to raw young fellows". The study of the ancient knowledge he shows, has given rise to hardly a single useful discovery, which fact is, for him, enough to condemn it.

Bacon's plan is: (1) to collect *reliable tested information*, especially by means of experiments.

(2) to classify this material by "*tables of invention*" so that all the instances of the phenomenon studied could be compared.

(3) By these tables he would arrive at minor generalisations, which we would call theorems or rules, and, by comparing these, he would rise to general *Scientific laws*.

(4) These Laws, when found, must confirm themselves by pointing out *new instances of the phenomenon studied*.

This is not very far from the modern scientific method, from which it differs, first, in that we use our judgment in selecting the questions we ask of nature, instead of collecting every fact that may possibly bear on the subject; and secondly, that we endeavour to collect *numerical* data, and handle them by *mathematical* reasoning.

Bacon considered that the purpose of science was not to

make up new theories about the nature and principles of things, but to lay firmer foundations of human power and greatness. So Science was to minister to the Arts—to medicine, navigation, industry of all kinds. This was almost a new conception, though there is a plea for it in the work of Roger Bacon more than three centuries before.

The work of Francis Bacon had the greatest influence, and whether we admire his ideas or not, we must recognise in him the founder of the school of English scientists that were the wonder of the seventeenth century.

On the continent René Descartes did something of the same work as Bacon. His *Discourse on Method* also purports to be an account of a new way of finding out the principles of nature. But while Bacon was a lawyer and believed in evidence, Descartes was a mathematician and believed in reasoning. Descartes believed that, just as Euclid *seemed* to have got his axioms out of his head and to have evolved from them the vast structure of geometry, so the scientist, merely by thinking, with only a minimum of observation of phenomena, would be able to arrive at axioms or principles from which long trains of reasoning would lead to an understanding of the whole course of nature. He was wrong, of course, erring far more in one direction than Bacon in the other. Bacon gave too much attention to gathering facts and little or none to mathematical reasoning about them: Descartes made precisely the opposite error. The practical scientists of the subsequent decades found out the happy mean; and such men as Boyle and Newton were able to treat Bacon's observations and experiments by Descartes' mathematical reasoning, with the greatest success.

Examples of the Science of the Early Seventeenth Century

WILLIAM GILBERT'S TERRELLA OR MAGNETIC MODEL OF THE EARTH

. . . Take then a strong loadstone, solid, of convenient size, uniform, hard, without flaw: on a lathe, such as is used in turning crystals and some precious stones, or on any like instrument (as the nature and toughness of the stone may require, for often it is worked only with difficulty), give the loadstone the form of a ball." . . . "To find, then, poles answering to the earth's poles, take in your hand the round stone, and lay on it a needle or a piece of iron wire: the ends of the wire move round their middle point, and suddenly come to a standstill. Now, with ochre or with chalk, mark where the wire lies still and sticks. Then move the middle or centre of the wire to another spot, and so to a third and a fourth, always marking the stone along the length of the wire where it stands still: the lines so marked will exhibit meridian circles, or circles like meridians on the stone or terrella: and manifestly they will all come together at the poles of the stone."

[Much of this was anticipated by Peter Peregrine.]

PERPETUAL MOTION BY MAGNETS IMPOSSIBLE

"Cardan writes that out of iron and loadstone may be constructed a perpetual-motion engine—not that he saw such a machine ever: he merely offers the idea as an opinion, and quotes from the report of Antonius de Fantis, of Treviso: such a machine he describes in Book IX, *De Rerum Varietate*. But the contrivers of such machines have but little practice in magnetic experiments. For no magnetic attraction can be greater (whatever art, whatever form of instrument you employ) than the force of retention; and objects that are conjoined, and that are near, are held with greater force than objects solicited and set in motion are made to move; and as we have already shown, this motion is a coition of both, not an attraction of one. Such an engine Petrus Peregrinus, centuries ago, either devised or delineated after he had got the idea from others; and Joannes Taysner published this, illustrating it with wretched figures, and copying word for word the theory of it. May the gods damn all such sham, pilfered, distorted works, which do but muddle the minds of students!"
(*On the Loadstone and Magnetic Bodies and on the Great Magnet the Earth.* William Gilbert of Colchester. Translation by P. F. Mottelay. Quaritch. 1893. The Latin original was published in 1600.)

STEVIN CONFUTES ARISTOTLE'S VIEWS ON FALLING BODIES

". . . The experiment against Aristotle is this: let us take (as I have done in company with the learned H. Jan Cornets de Groot, most diligent investigator of Nature's mysteries) two leaden balls, one ten times greater in weight than the other, which allow to fall together from the height of thirty feet upon a board or something from which a sound is clearly given out, and it shall appear that the lightest does not take ten times longer to fall than the heaviest, but that they fall so equally upon the board that both noises appear as a single sensation of sound. The same, in fact, also occurs with two bodies of equal size, but in ten-fold ratio of weight."

(*De Beghinselen des Waterwichts*, Simon Stevin. 1586.)

GALILEO STUDIES BODIES ROLLING ON AN INCLINED PLANE

"In a prisme" (*i.e. a beam of rectangular cross section*) "or Piece of Wood, about twelve yards long and half a yard* broad one way and three inches the other, we made upon the narrow Side or Edge a groove a little more than an inch wide: we shot it with the Grooving Plane very Straight, and to make it very smooth and sleek, we glued upon it a piece of Vellum polished and smoothed as exactly as can be possible: and in it we have let a brazen Ball, very hard, round and smooth, descend. Having placed the said Prisme Pendent (i.e. *inclined*) raising one of its ends above the Horizontal Plane a yard or two at pleasure, we have let the ball (as I said) descend along the Groove, observing, in the manner that I shall tell you presently, the Time which it spent in running it all, repeating the same observation again and again to assure ourselves of the Time, in which we never found any difference, no not so much as the tenth part of one beat of the Pulse. Having done, and precisely ordered this businesse, we made the same Ball to descend only the fourth part of the length of that Groove: and having measured the time of its descent, we alwaies found it to be punctually half the other. And then making trial of other parts,

* *Braccia*, in the original—about 2 feet.

examining one while the Time of the whole length with the Time of half the length, or with that of $\frac{2}{3}$, or of $\frac{3}{4}$, or, in brief, with any whatever other Division, by Experiments repeated near a hundred times, we alwaies found the Spaces to be to one another as the square of the Times. And this in all Inclinations of the Plane, that is, of the Groove in which the Ball was made to descend. In which we observed moreover, that the Times of the descents along sundry Inclinations did retain the same proportion to one another, exactly, which anon you will see assigned to them and demonstrated by the Author. And as to the measuring of the Time: we had a good big Bucket full of water hanged on high, which by a very small hole, pierced in the bottom, spirted, or, as we say, spin'd forth a small thread of Water, which we received with a small cup all the while that the Ball was descending the Groove, and in its parts: and then weighing from time to time the small parcels of water, in that manner gathered, in an exact pair of scales, the differences and proportions of their Weights gave justly the differences and proportions of the Times; and this with such exactnesse, that, as I said before, the trials being many and many times repeated, they never differed any considerable matter."

(Note (a) measurement, (b) repetition of experiment under various conditions, (c) deduction of a mathematical result.)

GALILEO WEIGHS AIR

". . . I have taken a pretty bigge Glasse Bottle with a narrow neck, and a Finger-stall of Leather fast about it, having in the top of the said Finger-stall inserted and fastened a Valve of Leather, by which with a Sirienge I have made passe into the bottle by force a great quantity of Air, of which, because it admits of great Condensation, it may take in two or three other Bottlesful over and above that which is naturally contained therein. Then I have in an exact Ballance very precisely weighed that Bottle with the Air compressed within it, adjusting the weight with small Sands. Afterwards the valve being opened and the air let out, that was violently contained in the Vessel, I have put it again into the Scales, and finding it notably

aleviated (i.e. lighter), I have by degrees taken so much Sand
from the other Scale, keeping it by it self, that the Ballance
hath at last stood in *Equilibrio* with the remaining counterpoise,
that is with the bottle. And here there is no question, but that
the weight of the reserved Sand is that of the Air that was
forceably driven into the bottle, and which is at last gone out
thence. But this Experiment hitherto assureth me of no more
but this, that the Air violently detained in the Vessel weigheth
as much as the reserved Sand, but how much the Air resolutely
and determinately weigheth in respect of the Water, or other
grave* matter, I do not as yet know, nor can I tell, unlesse I
measure the quantity of the Air compressed and for the discovery of this a Rule is necessary, which I have found may
be performed two manner of wayes. . . . The other way is
more expeditious, and it may be done with one Vessel only,
that is with the first accommodated after the manner before
directed, into which I will not that any other Air be put more
than that which naturally is found therein: but I will, that we
inject Water without suffering any Air to come out, which
being forced to yield to the super-venient Water must of
necessity be compressed: having gotten in, therefore, as much
water as possible, (but yet without great violence one cannot
get in three quarters of what the Bottle will hold) put it into
the Scales, and very carefully weigh it: which done, holding the
vessel with the neck upwards, open the Valve, letting out the
Air, of which there will precisely issue forth so much as there
is Water in the Bottle. The Air being gone out, put the Vessel
again into the Scales, which by the departure of the Air will
be found lightened, and abating from the opposite Scale the
superfluous weight, it shall give us the weight of as much Air
as there is Water in the Bottle."

(Both from *Dialogues concerning two new Sciences*, Galileo Galilei.
1638. Translated by T. Salusbury. 1661–5.)

HARVEY COMPUTES THE VOLUME OF BLOOD PASSING
THROUGH THE HEART

". . . But so that no one may say we are giving only words,
* Heavy.

and making specious assertions without foundation, and making innovations without just cause; there come up for confirmation three things from which, if granted, this truth necessarily follows and the matter is quite clear.

"First: that the blood is continually and without interruption being transmitted from the vena cava into the arteries by the beating of the heart and in such quantity that it could not be supplied by what was taken in, so much so that the whole mass (of blood) passes out of the heart in a short time. . . .

"Let us make a supposition (either by reasoning or experiment) as to the amount of blood the left ventricle holds when dilated (when it is full), whether two, or three, or one-and-a-half ounces: I have found two in the dead body. Let us in the same way assume that the heart, (when contracted) holds that amount less, by which it is smaller under those conditions, and by which the ventricle is then less capacious; and let us suppose that the above amount of blood is forced out of the ventricle into the aorta: . . . then one may reasonably conjecture that a fourth or fifth or sixth or at least an eighth part is sent into the artery. So we may assume that in man there is put forth from the heart half an ounce or three drachms or two drachms; which because of the closing of the valves, cannot flow back to the heart.

"In one half-hour the heart makes more than a thousand beats; and in some people and at some times, two, three or four thousand. Now multiply the drachms, and you will see that in one half-hour there is poured through the heart into the arteries, a thousand times either three drachms (or two drachms), that is five hundred ounces, or some other proportionate quantity of blood, which is a greater quantity than is found in the whole body."

(Translated from *Exercitiones Anatomicae De Motu Cordis, et Sanguinis Circulatione*, by William Harvey. First published, 1628.)

FRANCIS BACON ON THE SCIENTIFIC METHOD

Man, who is the servant and interpreter of nature, can act and understand no farther than he has, either in operation, or in contemplation, observed of the method and order of nature.

Neither the hand without instruments, nor the unassisted understanding, can do much; they both require helps to fit them for business; and as instruments of the hand, either serve to excite motion, or direct it; so the instruments of the mind either suggest to, or guard and preserve the understanding.

. . . Those who have treated the sciences were either empirics or rationalists. The empirics, like ants, only lay up stores, and use them; the rationalists, like spiders, spin webs out of themselves; but the bee takes a middle course, gathering her matter from the flowers of the field and garden, and digesting and preparing it by her native powers. In like manner, that is the true office and work of philosophy, which, not trusting too much to the faculties of the mind, does not lay up the matter, afforded by natural history and mechanical experience, entire or unfashioned, in the memory, but treasures it, after being first elaborated and digested in the understanding; and, therefore, we have a good ground of hope, from the close and strict union of the experimental and rational faculty, which have not hitherto been united.

. . . Again, in the very stock of mechanical experiments there is a great want of such as principally conduce to the information and the understanding. For the mechanic, being in no way concerned about the discovery of truth, applies his mind, and stretches out his hand, to nothing more than is subservient to his work; but we may then rationally expect to see the sciences farther advanced, when numerous experiments shall be received and adopted into natural history, which of themselves are useless, and tend only to the discovery of causes and Axioms; those being what we call experiments of light to distinguish them from experiments of profit. And they have this wonderful property, that they never deceive or frustrate the expectation: for being used, not in order to effect any work, but for disclosing natural causes, in certain particulars; let them fall which way they will, they equally answer the intention, and solve the question.

And not only a larger stock of experiments is to be sought, and procured, of a different kind from what has hitherto appeared, but also a quite different method, order, and procedure, is to be introduced, for continuing and advancing experience itself; for vague experience, that pursues nothing but itself, is, as was before observed, a mere groping about in the dark, and rather amazes mankind, than informs them. But when experience shall be conducted by certain laws, orderly and consequentially, we may have better hopes of the sciences.

And when a proper quantity of suitable materials, for such a natural and experimental philosophy, as is required for the work of the understanding, or the business of philosophy, shall be provided and got ready, yet the understanding is no way qualified to act upon these materials spontaneously, and by memory, no more than a man should expect to make the calculations for an almanack,* by the bare strength of his memory. Yet contemplation has hitherto been more employed in invention than writing, nor is experience yet made learned. But no invention should be allowed, or trusted, except in writing. And when this comes into use, so that experience may be made a matter of learning and science, we may then have better hopes.

Again, the number, or, as it were, the army of particulars, being so large, scattered and confused, as to distract and confound the mind, little good can be expected from the skirmishes and sallies of the understanding, unless it be fitted, and brought close to them, by means of proper, well-disposed and actuating tables of invention, containing such things as belong to the subject of every enquiry; and unless the mind be applied to receive the prepared and digested assistance they afford.

And even when a stock of particulars is exactly and orderly placed before us, we must not immediately pass on to the enquiry and discovery of new particulars or works at least if this be done, we must not dwell upon it. We deny not, that after all the experiments of every art shall be collected, digested. and brought to the knowledge and judgment of a single person, many new discoveries may be made, for the use and advantage of life, through the translation of the experiments of one art into another, by means of what we call learned

* i.e. a table of risings and settings of planets, etc.

experience; yet less hope is to be conceived hereof and a much greater of a new light of Axioms, drawn regularly, and in a certain manner, from those particulars, so that such Axioms may again point out, and lead to new particulars. For the way lies not through a plain, but through mountains and valleys, first ascending the Axioms, and then descending to works.

But the understanding must not be allowed to leap, or fly from particulars, to remote, or the most general kind of Axioms, at once (such as are called the principles of arts and things), and so prove, and draw out middle Axioms, according to the established truth of the former, as has hitherto been done by a natural sally of the understanding, which is naturally inclined this way, and has been long trained and accustomed to it, by the use of those demonstrations which proceed upon syllogism. But we may conceive good hopes of the sciences when, by continued steps, like real stairs, uninterrupted or broken, men shall ascend from particulars to lesser Axioms, and so on to middle ones; from these again to higher; and, lastly, to the most general of all.

. . . And, first, the introduction of noble inventions seems to hold by far the most excellent place among all human actions. And this was the judgment of antiquity, which attributed divine honours to inventors, but conferred only heroical honours upon those who deserved well in civil affairs, such as the founders of empires, legislators, and deliverers of their country.

. . . If anyone, in the last place should object that the arts and sciences may be wrested, and turned to evil purposes, or sin, luxury, etc., this can have little weight, because it may be said of all the best things in the world, such as great capacity, courage, strength, beauty, riches, light itself, etc. Let but mankind recover their right over nature, which was given them by the Divine Being, let them be well provided of materials, and rectified reason and sound religion will direct the use.

(*Novum Organum*, Francis Bacon. Part I.)

DESCARTES' RULES FOR INVESTIGATION

The first was never to take anything as true that I did not know evidently to be so; that is to say, carefully to avoid haste and prejudice, and to include nothing more in my judgements than that which should present itself so clearly and distinctly to my mind, that I should have no occasion to doubt of it.

The second, to divide each of the difficulties which I might examine into as many portions as should be possible and should be necessary, the better to resolve them.

The third, to conduct my thinking in an orderly manner, beginning with the objects most simple and most easy to understand, in order to rise little by little, as if by steps, up to the knowledge of the most complex; supposing moreover that there is an order even among those that do not proceed naturally one from the other.

And the last, always to make enumerations so complete and reviews so general, that I should be certain of having omitted nothing.

Those long chains of reasoning, all simple and easy, of which the geometers are accustomed to avail themselves in order to arrive at the most difficult demonstrations, have caused me to imagine that all things that come within the knowledge of man follow upon each other in the same fashion; and that, provided only that nothing is taken for true that is not so and the necessary order is preserved for the deduction of one from another, there can be none of them so distant that it cannot be arrived at, nor so hidden that it cannot be discovered.

(*Discours de la Méthode pour bien conduire sa Raison et chercher la vérité dans les sciences.* René Descartes. Second Part. 1637.)

CHAPTER SEVEN

Man Learns the Nature of the Solar System

Astronomical teaching before 1610

The Copernican theory was published in 1543, but it gained ground very slowly and even at the beginning of the seventeenth century its adherents were to be reckoned only in dozens, and the whole mass of the learned world adhered to the views that had been current from antiquity. Yet by the end of that century, the ancient astronomy was entirely discredited; and it is our task to show how that came about.

What were the fundamental views of the old astronomy, based on Aristotle?

The universe was a closed sphere. At the centre was the motionless earth, around which there revolved successive and ever wider spheres carrying the moon, sun, planets and fixed stars respectively. Everything below, i.e. within, the sphere of the moon was *terrestrial*, made of earth, air, fire and water, and was subject to change and decay. The natural motion of terrestrial bodies, i.e. that which they pursued if not restrained or impelled by violence, was up (fire) or down (earth); *up* and *down* meaning *away from* and *toward* the centre of the universe. Outside the sphere of the moon was the *celestial* region. Everything there was made of a fifth and nobler element, imperishable, eternal, immutable. Nothing above the moon could come to be or pass away. Objects which obviously did so, such as comets and shooting stars, were not considered to be in the celestial region but in the upper air. The natural motions of heavenly bodies were circular, uniform and unchangeable; and, being natural, required no explanation. The complex motions which the heavenly bodies actually displayed were explained as the resultant of several simultaneous uniform circular motions.

The Ptolemaic theory, as held in the sixteenth century, admitted all this, but it also allowed of motions in circles whose centres were not at the centre of the universe, and it also required a circular motion which was not strictly uniform; this latter was Copernicus's chief objection to it.

The Aristotelean view of astronomy was not, of course, a part of religious dogma, but naturally all the great theologians and philosophers had held it, for the same reason that they now hold the Newtonian view, namely, because it was the one accepted by the experts of their day. Since it is difficult to discuss the origin and nature of the universe without bringing in astronomical views, the works of the great mediæval theologians and philosophers, such as St. Thomas Aquinas, contain a great deal of Aristotelean astronomy, and the respect that was rightly given to their philosophy and theology was apt to be extended to their astronomy; nor was a welcome given to those who sought to overthrow it. The University Professors, too, were Aristotle-experts, one and all, and naturally disliked the prospect of having to discard their methods and learn astronomy afresh from the experimental scientists. It is not surprising that these influential adversaries gave the new astronomical views a stormy passage.

The two great systems of the world are usually referred to as respectively *geocentric* (having the earth at the centre of the planetary system), and *heliocentric* (having the sun at the centre of that system), although after the time of Kepler it was clear that the sun was not at the geometrical centre of the planetary orbits. The geocentric systems include both the Aristotelean and Ptolemaic: while the heliocentric includes the Copernican system and all its modern modifications.

Evidence for the Heliocentric System

Copernicus produced no *compelling* evidence that his system was the true one. He showed that it was simpler and more orderly than the geocentric, and that it agreed quite as well with the observations. But the agreement was far from perfect, because, in fact, the planetary orbits are not circles but ellipses, and also because Copernicus made use of many old incorrect observations that no theory could possibly have accounted for. During the hundred and fifty years which elapsed between the publication of the theory of Copernicus and its universal acceptance there were three great events made people believe in it:

(1) The astronomical telescope was invented and disclosed some things which confirmed the heliocentric view (Galileo, 1610).

(2) The Copernican theory was modified by adopting elliptical orbits and was then shown to agree with the observed motions within the limits of error of observation. (Kepler, 1609–19.)

(3) It was shown by Newton in 1687 that, if the heavenly bodies obeyed the same laws of motion as do bodies on earth, the whole Copernican system, as modified by Kepler, could be deduced from those laws; and this left no reasonable grounds for doubt that the heliocentric system, with planets revolving in elliptical orbits, was the true one.

The astronomical Work of Galileo

We have already shown how Galileo disproved Aristotle's laws of motion, and a great deal of the older astronomy was based on those laws. In the years between 1600 and 1610 Galileo gave some attention to the phenomena of comets and new stars, *novae*, of which two very conspicuous examples had occurred in 1572 and 1604. He showed that these were further away from us than the moon, that is to say that they were outside its 'sphere' and in the 'celestial' region. Therefore they were celestial bodies and the appearance and subsequent fading out of these stars were changes in celestial bodies, and so Aristotle's notion of their immutability and freedom from generation or destruction was shown to be false. But astronomical observations were still inaccurate, and Galileo's adversaries found plenty of evidence in support of the view that these bodies were really below the moon, and were mere atmospheric phenomena.

In 1609 Galileo heard that a Dutchman had made a telescope, and he at once figured out how it might be done, and set to work to make one. He had to make it himself and to grind his own lenses, so naturally the earlier ones were not very good.

FIG. 33.—Galileo in middle life. The cherub on the right holds
a telescope, and that on the left a proportional compass, also
invented by Galileo. (From Sherwood Taylor's *Galileo and the
Freedom of Thought*, by courtesy of Messrs. Watts and the
Trustees of the British Museum.)

But he had great mechanical skill, and within a few months was able to direct the first telescope to the sky, and reap a harvest of discovery. A great part of the rest of his life was spent in applying these discoveries and other evidence to the attempt to support or prove the Copernican view.

We may tabulate these great discoveries.

(1) Galileo interpreted his telescopic view of the moon as showing it to be a rugged rocky body like the earth, not a smooth lucid globe of a pure fifth element. This seemed to argue that at least one heavenly body was of the same stuff as the earth. Not only did this seem to disprove the existence of Aristotle's celestial element, but it led to the argument that if the moon, which was a planet, was of the same stuff as the earth, it was reasonable to suppose that the earth was a planet, as Copernicus had said and his opponents had denied.

(2) Venus showed phases like the moon; so presumably it was also a solid opaque body like the moon, and did not shine by its own light.

(3) The telescope showed spots on the sun,* and these came into being, shifted and disappeared: so the sun, noblest of all celestial beings, was not immutable.

(4) The telescope showed Jupiter to be surrounded by four little satellites which rotated about him in very short periods. Here was an ocular demonstration of what Copernicus had asserted of the earth and moon, and it showed, moreover, that the smaller bodies revolved about the bigger, which agreed with the heliocentric system, but not with the geocentric.

(5) The telescope showed a host of stars far exceeding in number those visible to the eye. It was evident that all these stars had been unknown to the ancients, whose view of the universe was thereby demonstrated to be incomplete.

---

* Galileo was probably the first to see sunspots though he was not the first to publish the discovery.

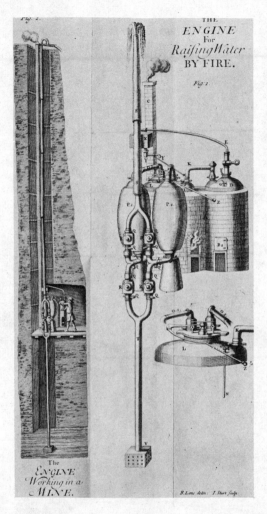

PLATE XI
Savery's pumping engine, invented in 1698.

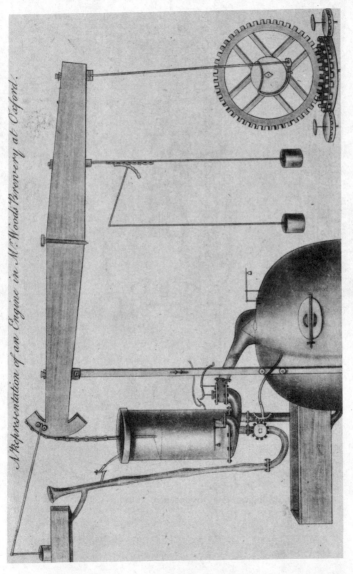

A Representation of an Engine in Mr Wood's Brewery at Oxford.

PLATE XII

Engine of Newcomen type (c. 1780) used to drive a wheel, presumably for some kind of stirring operation. Left to right: Pump to raise cooling-water, pipe to carry this to the cylinder, jet to cool top of piston and seal leakages: cylinder and piston: rod with projections designed to open and shut the cold-water valve and steam valve: boiler with safety valve: rocking beam (support not shown): counterpoise and pump (?): connecting rod and crank.

Galileo, when he wrote his great book the *Dialogues Concerning the Two Principal Systems of the World* (1632), advanced all this evidence and more, and utterly demolished the ancient astronomy. But he was not content to break down the old views, and his great desire was to establish the Copernican system in their place. But in trying to find proofs of the heliocentric system he over-reached himself. He attached special importance to a pet theory that the tides were caused by the earth's rotation in a manner which was not only fallacious, but did not even agree with the well-known fact that there are two high tides a day and that these may occur at any hour of the day, and he also produced another fallacious proof from the apparent path of sunspots. These fallacies diminished the force of his other good arguments, but in spite of them he converted most of the scientific world to the heliocentric view.

A word must be said of Galileo's unhappy conflict with the Church. He had made himself many enemies among the university professors (who were almost all ecclesiastics), and he was several times denounced to the eccesiastical authorities. For some time they supported Galileo, but in 1616 a judicial opinion was given that the heliocentric system was heretical. It seems that this decision was not only unwise, but had no satisfactory grounds in theology. Galileo was ordered to keep silent on the subject of heliocentric astronomy and agreed to do so; but some years later, counting on the expected favour of the new Pope, Urban VIII, he published his *Dialogues* in 1632, and was summoned before the Inquisition, forced to recant, and confined to his house for the rest of his life. These events did no credit to the ecclesiastics who were responsible, nor do Galileo's shifty methods of trying to evade the order of 1616 in spirit while feigning obedience to it in letter make us think the better of him.

This is the only case of real conflict between the Catholic Church and Science, for the few other men of science who were punished by the Inquisition or were excommunicated, were so treated on account of heretical theology, or a suspicion of witchcraft, sometimes well founded.

Kepler's Laws

Johann Kepler (1571–1630) is an interesting figure. He was
deeply impressed with the order and harmony of the universe
and he felt quite sure that, if he could but discover them, there
were simple mathematical formulæ or relationships which would
express the connection between such apparently unrelated
quantities as the distances of the planets and the times of their
revolutions. He would not be content with anything but the
most precise agreement between his formulæ and the observed
facts; and it was this extreme persistence in looking for perfection that brought him his success. His main work was the
study of the motion of Mars. He began as Tycho Brahe's
assistant and inherited his observations, but very few of them

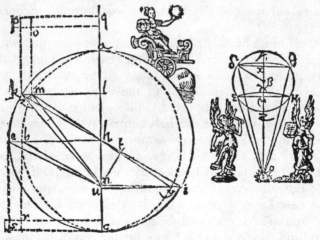

Fig. 34.—Kepler's figure showing the elliptical figure of the
orbit of Mars (dotted). Note the triumphal chariot.

proved to be of use to him. He himself made accurate observations over many years, and being a Copernican, he expressed
them in terms of revolutions of Mars about the sun, not about
the earth, and it was this that made his success possible. His
problem was to find some path that accounted for the fact that
Mars moved sometimes more swiftly, sometimes more slowly.
Circular motion would not do, whether concentric or eccentric:

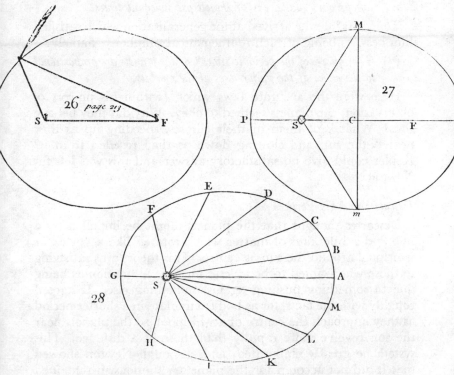

FIG. 35.—Illustrating Kepler's Laws. (From Long's *Astronomony.* 1700.)
Top left. Drawing an ellipse. The pins are at the two foci.
Top right. The sun in one focus of the elliptical orbit of the planet. PA is the major axis.
Below. The areas GSF, FSE, ESD etc. are equal in area. The planet then traverses the portions of its orbit GF, FE, ED etc. in equal times.

he then tried various kinds of 'oval' paths, which brought him within 4′ of the correct positions. Even then he was unsatisfied, though the agreement was so close that Copernicus could not even have measured so small a discrepancy. So he finally was led to the idea of an elliptical path, and in 1609 he formulated the following two laws which appeared to hold exactly:

(1) *The planets describe ellipses about the sun, the sun being in one focus.*

(2) *The planets move so that the line joining the sun to the planet sweeps out equal areas of the ellipse in equal times.*

Ten years later he arrived at the generalisation which satisfied him best, as being of such mathematical simplicity, namely:

(3) *The squares of the periodic times* of the planets are proportional to the cubes of the major axes of their orbits.*

Here were the accurate laws which (within his powers of observation) the planets seemed to obey. Why did they behave thus? What kept them in their ellipses, speeding up as they neared the sun and slowing down as they receded from it? Kepler could give no satisfactory answer and this was left for Newton.

Descartes and his Vortices

Descartes thought that the planets might be impelled by a fine and subtle kind of matter which rotated like a vortex or whirlpool around the sun as centre. This theory was satisfying to those who wanted to be able to imagine all motion as being due to something pushing against something else. It agreed roughly with the facts, for as bodies circulate in a shorter period as they approach the centre of a whirlpool, so the planets near the sun revolve more rapidly than those at a distance. This system he greatly elaborated, and, although Newton showed that it did not account for the planetary motions, it remained the prevailing one on the Continent until 1730–40.

The work of Isaac Newton

There is no doubt as to what Newton did, but it is not very clear what mental processes led him to it. In the year 1666, when he was only 24, he arrived at his fundamental notion that every particle of matter attracts every other particle, and he thought that the attraction probably varied directly as the product of the masses and inversely as the square of the distance between them. He could not verify this by an experiment because the force of gravitation between two bodies small

---

* The periodic time of the planet is the time of one revolution about the sun: the major axis of an ellipse is the part of the straight line passing through its foci which is cut off by its circumference (e.g. the line PA in Fig. 35).

FIG. 36.—The vortices of Descartes:
planets are carried in the whirlpool
of subtle matter about the sun S.

enough to be handled is so minute. It struck him that the force
that makes bodies fall (i.e. that attracts objects to the earth)
being apparently undiminished at the tops of high mountains,
might extend indefinitely—as far indeed as the moon. This gave
him a chance to test his theory. He knew that on earth a stone
fell 16 feet in the first second. Then if he supposed (what he
could not yet prove) that the stone was attracted by the earth
as if the latter's attractive power were concentrated at its
centre, he could say that the moon ought to be falling in every
second a distance given by

$$\frac{16 \times (radius\ of\ earth)^2}{(distance\ of\ moon's\ centre\ from\ earth's\ centre)^2}$$

Now the moon actually is falling. If the earth's attraction
were suddenly removed the moon would travel on in a straight

line MT (Fig. 37), so in the time the moon takes to travel
from M to M¹ it has been forced to approach the earth by a
distance TM¹, i.e. to fall that distance. Knowing the circumference of the moon's orbit and the period of rotation (1 month),
it is easy to calculate the speed of fall. Newton did so and it
came to about 15% less than his former calculation indicated.
This, though 'it answered pretty nearly', was not good enough
for him and he laid the calculations by. What was the source
of the discrepancy? Simply that Newton adopted for the
radius of the earth the current value, which had been in use
since the time of Ptolemy and was a good deal too small. Some
twelve years later, better measurements were made, and
Newton thus discovered that his former calculation had been
based on incorrect observations. He waited three years before
recalculating the results, probably because he was unable to
prove, as yet, that a body at the surface of e.g. the earth was
attracted by it with the same force as would be exerted by its
whole mass concentrated at the centre. His calculations then
came out correct—thus giving very strong evidence for his
theory. He followed it up and showed that if the force that
kept the planet in their orbits was gravitational attraction
varying inversely as the square of the distance between attracting bodies, those orbits must be ellipses.

But Newton had no interest in publishing his discoveries,
for he hated the controversies to which publication generally
led; so he let this gigantic achievement lie by him for five years
and even lost the proof. Only when the astronomer Halley
asked him what would be the path of a planet moving under
the attraction of a force which varied according to an inverse-
square law, did he reply "An ellipse"; and only at Halley's
persuasion did he undertake to publish his theory. Once he
was moved to work, Newton's power was astonishing. His
great book *Philosophiæ Naturalis Principia Mathematica* (1687)
is perhaps the most powerful and original piece of scientific
reasoning ever published. Newton created the science of
dynamics, defining force, momentum, etc., for the first time.
He invented most of the mathematics needed for the work;
the inverse-square law, the principle of gravitation and the

idea of applying it to the heavens were his. Starting with these simple principles, he builds up, by a process as rigid as that of propositions in geometry, the proof of the whole motions of the solar system as they were then known. He disproved Descartes' theory of vortices, and explained the precession of the equinoxes and the theory of the tides.

Newton's Researches on Colours

In January 1666 Newton made one of his greatest discoveries, the theory of colours, which, since it led him to the invention of the reflecting telescope, may be mentioned here. He passed a beam of light through a prism and found it lengthened out into a multicoloured strip. He did not announce this till six years later (1672) when he communicated it to the Royal Society. In this paper he concluded then that white light was compound, consisting of rays which were refracted to different extents, and that each colour was not a "qualification" of light, but was a property of rays of a particular 'refrangibility'. Here was a very startling expression of a quality in terms of a quantity. Lenses were well known to give images with colour fringes and this was already the chief obstacle to clear vision with telescopes and microscopes. To Newton it seemed that this fault was inseparable from a refracting instrument, and so he set to work to make a telescope using a mirror instead of a lens. He spent a great deal of time on compounding alloys suitable for the metal mirrors. The reflecting telescope came into its own in the second half of the eighteenth century, but after achromatic refracting telescopes were perfected the reflector suffered a period of eclipse, but has now regained its position for telescopes of the highest powers.

Newton's theories of the nature of light were very influential. The wave-theory of light had been put forward by Huyghens, but Newton could not accept it because he did not believe that a wave-motion could cast sharp shadows, and supposed that it would bend round the corner of an obstacle. None the less the colours of thin films (e.g. Newton's rings) required some periodic property to explain them. So he adopted the theory that light consisted of particles which could arouse a vibration.

He tells us that ". . . it is to be supposed that the aether is a vibrating medium like air, only the vibrations far more swift and minute. . . . I suppose light is neither aether nor its vibrating motion, but something of a different kind propagated from lucid bodies."

Newton's successors

Newton gave the world the main outlines of the plan of the solar system and the principles by which it was to be understood; for the next 150 years the astronomical and mathematical world did little else but develop what he had started. The chief astronomical problem was to calculate the extent to which the elliptical motion of the planets as set out by Kepler's laws is modified by the small attractions of the planets on each other. This requires very difficult and tedious mathematical analysis. The various small irregularities of the planets were explained one by one. The final triumph came in 1846, when Adams and Leverrier independently deduced the existence and place of a planet more distant than Uranus from the unexplained irregularities of the motion of the latter. The discovery of this planet, Neptune, virtually concluded the story of the mapping of the solar system, though many small bodies, satellites, asteroids, and recently the distant planet Pluto, have since been added.

Examples of Seventeenth-Century Astronomy

GALILEO MAKES HIS ASTRONOMICAL TELESCOPE

About ten months ago a rumour came to our ears, that an optical instrument had been made by a certain Dutchman, by aid of which visible objects, though very distant from the eye of the observer, were seen distinctly as if near to him. . . . The same was confirmed to me a few days later by a letter written from Paris by the noble Frenchman Jacob Badovere of Paris, and so I set myself to inquire into the principles and means by which I might be able to arrive at the discovery of a similar instrument, in which a little later I succeeded by a study of the theory of Refraction. And so I prepared a leaden tube, at the ends of which I fixed two glass lenses both plane on one face, one of them being spherically convex, the other concave on the other face. Then applying my eye to the concave end, I saw objects as quite large and near, for they appeared three times nearer and nine times larger* than before. . . . At length, sparing no labour and no expense, I succeeded in making an instrument so excellent that things seen through it appeared nearly a thousand times larger and more than thirty times nearer, than if viewed by natural vision alone.

(From *Sidereus Nuncius*, or *The Starry Messenger*, Galileo Galilei. 1610.)

THE MOUNTAINS ON THE MOON

. . . from my observations . . . I have been led to understand clearly that the surface of the Moon is not polished, even, and very exactly spherical, as a large school of Philosophers consider her and the other heavenly bodies to be, but on the contrary is uneven, rough, full of hollows and protuberances, not otherwise than is the face of the earth itself which is everywhere marked out by peaks of mountains and the hollows of valleys.

The appearances from which one may deduce this are as follows: On the fourth or fifth day after new moon, when the Moon displays itself to us with bright horns, the boundary which separates the part still dark from that which is bright does not extend evenly in the form of an elliptical curve, as would happen with a perfectly spherical solid, but is marked out by an uneven, rough and decidedly wavy line, as the attached figure shows. For several bright excrescences, as it were, extend beyond the boundary of light and darkness into the dark part and on the other hand dark portions enter into the light part. Indeed, a great number of small darkish spots, altogether separate from the dark part are scattered about almost the whole part which is then flooded with the Sun's light. . . . And I have noticed that these small spots always and in every case agree in having the dark part on the same

* In area.

side as the Sun's position and on the side opposite to the Sun they have brighter boundaries, as if they were crowned with shining peaks. But we have a quite similar appearance on Earth about sunrise, when we see the valleys not yet flooded with light but the mountains surrounding them on the side opposite to the sun already blazing with his splendour; and just as the shadows in the hollow places of the earth diminish as the sun grows higher, so also these spots on the Moon, lose their darkness as the bright portion grows larger. . . .

(From the work cited above.)

(This phenomenon is readily observable with a field-glass or small telescope, which will give a good idea of what Galileo saw. The definition of his telescope must have been poor, and probably an x8 field-glass shows us as much as he could have seen.)

GALILEO CONFUTES THE ARGUMENTS USED AGAINST THE COPERNICAN SYSTEM

(From the time of Ptolemy, text-books of astronomy reproduced stereotyped arguments which were designed to prove that the earth is stationary. Here is an example of such a 'proof' taken from the *Epitome Astronomiae* of Michael Mæstlin, published in 1582. Mæstlin must be remembered with gratitude for having persuaded the youthful Kepler, who had leanings to literary studies, to take up astronomy.)

Proof that the Earth is not Moved with a Circular Motion

If the Earth were carried about from East to West in a circular fashion, the daily sunrisings and settings would not occur nor would there be the alternation of days and nights but there would be always noon in one place, in another always dawn, etc.*

For if it were moved with any other circular motion whatever, the clouds would be seen to fly always and only in the opposite direction. Heavy bodies projected vertically upward would never fall back on to the same spot. The same heavy bodies

* This assumes that the Sun moves round the Earth, while the Earth rotates with the same angular velocity.

also when falling back would not drop perpendicularly on to the places beneath them, for these would be carried from under them by the swiftness of the Earth's motion; also animals and buildings, shaken by this violence of motion, would fall down. Furthermore all the parts of the earth (at any rate if the earth be supposed to undergo a daily rotation) would be scattered abroad by its irresistible swiftness, and the whole earth would have by now been consumed and fallen outwards into the sky (which is absurd).

Lastly since Nature has granted to a simple body only one simple motion, and motion in a straight line belongs to the Earth (as is to be seen in its parts*) therefore it is impossible to justify a circular motion for it.

Therefore the Earth remains immovable, set in its place, which is the centre of the universe.

(Here is a part of the argument by which Galileo refutes the assertion that a stone would not fall vertically from the top of a tower if the earth were rotating.)

Salviati . . . "The error of Aristotle, Ptolemy, Tycho, yourself, and all the rest, is grounded upon that fixed and strong persuasion, that the Earth standeth still, which you have not judgement nor power to depose, no, not when you have a desire to argue of that which would ensue, pre-supposing the Earth to move. And thus, in the other argument, not considering that whil'st the Stone is upon the Tower, it doth, as to moving or not, the same that the Terrestrial Globe doth, because you have concluded with your self, that the Earth stands still, you always discourse touching the fall of the stone, as it were to depart from rest: whereas it behoveth to say, that if the earth standeth still, the stone departeth from rest, and descendeth perpendicularly; but if the Earth do move, the stone likewise moveth with the like velocity, nor doth it depart from rest, but from a motion equal to that of the Earth, whereunto it intermixeth the supervenient motion of descent, and of those two composeth a third which is transversal or sideways."

Simplicio. "But for God's Sake, if it move transversely, how is it that I behold it to move directly and perpendicularly? . . ."

* Which fall vertically in a straight line.

Salviati, "In respect to the Earth, to the Tower, and to our selves, which all as one piece move with the diurnal motion together with the stone, the diurnal motion is as if it had never been . . ."

Sagredo. "Now I do remember a certain conceipt that came one day into my fancy, whilst I sailed in my voyage to Aleppo. . . . If the neb of a writing-pen that I carried along with me in the ship, through all my navigation from Venice to Scanderon,* had had a facultie of leaving visible marks of its whole voyage, what signs, what marks, what lines would it have left?"

Simplicio. "It would have left a line distended from Venice thither, not perfectly streight or to say better, distended in a perfect arch of a circle, but in some places more, in some less curved, according as the vessel had gone more or less fluctuating; but this . . . without any considerable error, might have been called the part of a perfect arch."

Sagredo. "If a Painter then at our launching from the Port had begun to design upon a Paper with that pen, and had continued his work till he came to Scanderon, he would have been able to have taken by its motion a perfect draught of all those figures perfectly interwoven and shadowed in on several sides with countreys, buildings, living creatures, and other things; albeit all the true, real, and essential motion traced out by the neb of that pen, would have been no other than a very long but simple line; and as to the proper operation of the Painter, he would have delineated the same to a hair, if the ship had stood still. That therefore of the huge long motion of the pen there doth remain no other marks, than those tracks drawn upon the paper, the reason thereof is because the grand motion from Venice to Scanderon, was common to the paper, the pen, and all that was in the ship; but the petty motions forwards and backwards, to the right, to the left, communicated by the fingers of the Painter into the pen, and not to the paper, as being peculiar thereunto, might leave marks of itself upon the paper, which did not move with that motion. Thus it is likewise, that the Earth moving, the motion of the stone in descending downwards was really a long tract of many hundreds

* Alexandretta.

and thousands of yards, and if it could have been able to have delineated in a calm air or other superficies, the track of its course, it would have left behind a huge long transverse line. But that part of all this motion which is common to the stone, the Tower, and ourselves is imperceptible to us, and as if it had never been, and that part onely remaineth observable, of which neither the Tower nor we are partakers, which is in fine, that wherewith the stone falling measureth the Tower."

(*Dialogues concerning the two principal systems of the World*, by Galileo Galilei. In *Mathematical Collections and Translations*. T. Salusbury. 1661.)

WHAT NEWTON DID AT THE AGE OF TWENTY-THREE

In the beginning of the year 1665 I found the method for approximating series and the rule for reducing any dignity [power] of any binomial to such a series [i.e. the binomial theorem]. The same year in May I found the method of tangents of Gregory and Slusius, and in November had [i.e. discovered] the direct method of Fluxions [i.e. the elements of the differential calculus], and the next year in January had the Theory of Colours, and in May following I had entrance into the inverse method of Fluxions [i.e. integral calculus], and in the same year I began to think of gravity extending to the orb of the Moon . . . and having thereby compared the force requisite to keep the Moon in her orb with the force of gravity at the surface of the earth, and found them to answer pretty nearly. All this was in the two years of 1665 and 1666, for in those years I was in the prime of my age for invention, and minded mathematics and Philosophy more than at any time since.

(MS of Newton cited in D.N.B. XIV. 371.)

NEWTON'S ACCOUNT OF KEPLER'S THIRD LAW

Phenomenon IV

That the fixed Stars being at rest, the periodic times of the five primary Planets, and (whether of the Sun about the Earth, or) of *the Earth about the Sun, are in the sesquiduplicate* proportion of their mean distances from the Sun.*

This proportion, first observ'd by *Kepler,* is now receiv'd by all astronomers. For the periodic times are the same, and the dimensions of the orbits are the same, whether the Sun revolves about the Earth, or the Earth about the Sun. And as to the measures of the periodic times, all astronomers are agreed about them. But for the dimensions of the orbits, *Kepler* and *Bullialdus,* above all others, have determin'd them from observations with the greatest accuracy: and the mean distances corresponding to the periodic times, differ but insensibly from those which they have assign'd, and for the most part fall in between them; as we may see from the following Table.

The periodic times, with respect to the fixed Stars, of the Planets and Earth revolving about the Sun, in days and decimal parts of a day.

♄	♃	♂	♁	♀	☿
10759,275	4332,514	686,9785	365,2565	224,6176	87,9692

The mean distances of the Planets and of the Earth from the Sun.

	♄	♃	♂
According to Kepler	951000	519650	152350
To Bullialdus	954198	522520	152350
To the periodic Times	954006	520096	152369

	♁	♀	☿
According to Kepler	100000	72400	38806
To Bullialdus	100000	72398	38585
To the periodic Times	100000	72333	38710

(*The Mathematical Principles of Natural Philosophy.* Sir Isaac Newton. Tr. Motte 1729. Vol. II. p. 210–11.)

NEWTON'S PROOF THAT THE GRAVITY IS THE FORCE THAT KEEPS THE MOON IN HER ORBIT

BOOK III. PROPOSITION IV. THEOREM IV.

That the moon gravitates towards the earth and by the force of gravity is continually drawn off from a rectilinear motion, and retained in its orbit.

(See Fig. 37)

---

* If *t* and *a* are in sesquiduplicate proportion, *t* varies as $a^{\frac{3}{2}}$.

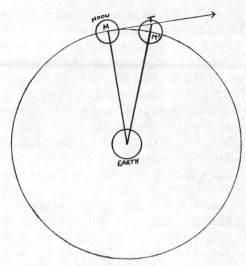

FIG. 37.—Illustrating Newton's proposition.

The mean distance of the moon from the earth in the syzygies* in semi-diameters of the earth, is, according to *Ptolemy* and most astronomers 59, according to *Vendelin* and *Huygens* 60; to *Copernicus* 60 1/3; to *Street* 60 2/5; and to *Tycho* 56½ . . .†

Let us assume the mean distance of 60 semi-diameters of the earth in the syzygies; and suppose one revolution of the moon, in respect of the fixed stars, to be completed in 27d. 7h. 43′, as astronomers have determined; and the circumference of the earth to amount to 123249600 *Paris* feet, as the *French* have found by mensuration. And now if we imagine the moon, deprived of all motion, to be let go, so as to descend towards the earth with the impulse of all that force by which (by cor. prop. 3.) it is retained in its orb, it will, in the space of one minute of time, describe in its fall 15 1/12 *Paris* feet. This we gather by a calculus, founded either upon prop. 36, book 1. or (which comes to the same thing) upon cor. 9, prop. 4 of the same book. For the versed sine of that arc,‡ which the moon,

---

* i.e. when earth, sun and moon are in line.

† After correcting Tycho's incorrect allowance for refraction, 60¼.

‡ The versed sine is the line TM′ in figure.

in the space of one minute of time, would by its mean motion describe at the distance of 60 semi-diameters of the earth, is nearly 15 1/12 *Paris* feet, or more accurately 15 feet 1 inch, and 1 line 4/9. Wherefore, since that force, in approaching the earth, increases in the reciprocal duplicate proportion of the distance, and, upon that account, at the surface of the earth, is 60 × 60 times greater than at the moon, a body in our regions, falling with that force, ought, in the space of one minute of time, to describe 60 × 60 × 15 1/12 *Paris* feet; and in the space of one second of time to describe 15 1/12 of those feet;* or more accurately 15 feet 1 inch and 1 line 4/9. And with this very force we actually find that bodies here upon earth do really descend;† for a pendulum oscillating seconds in the latitude of Paris will be 3 Paris feet and 8 lines ½ in length, as Mr. Huygens has observed. And the space which a heavy body describes by falling in one second of time is to half the length of this pendulum in the duplicate ratio‡ of the circumference of a circle to its diameter§ and is therefore 15 Paris feet 1 inch 1 line 7/9. And therefore the force by which the moon is retained in its orbit becomes, at the very surface of the earth, equal to the force of gravity which we observe in heavy bodies there. And therefore (by rule 1 and 2)‖ the force by which the moon is retained in its orbit is that very same force which we commonly call gravity, for were gravity another force different from that, then bodies descending to the earth with the joint impulse of

* Distance fallen = ½ acceleration × square of time, therefore if time is decreased from 60 seconds to 1 second, distance fallen decreased in the proportion of 60 × 60 : 1.

† Freely falling bodies could not be accurately timed in the seventeenth century, so Newton calculates the rate of fall from the rate of the seconds pendulum.

‡ i.e. "as the square of the ratio . . ."

§ This is readily deduced from the formula connecting the time of swing and length of a pendulum $t = 2\pi\sqrt{\dfrac{l}{g}}$: t representing the time of a double swing, 2 secs. for a seconds pendulum. The Paris foot was longer than the English.

‖ The substance of these rules is:
  (1) We are to admit no more causes of natural things than such as are both true and sufficient to explain the appearances.
  (2) . . . to the same natural effects, we must, as far as possible, assign the same natural causes.

both forces would fall with a double velocity, and in the space of one second of time would describe 30 1/6 *Paris* feet; altogether against experience.

(From *Principia Mathematica*. Sir Isaac Newton. 1687. Trans. Motte. 1803 ed.)

This whole proof requires no more than elementary geometry and mechanics and the student is recommended to work it out in detail.

CHAPTER EIGHT

The Mechanical Philosophy

Science, Abstraction and the Real World

All science in the intellect; and things, if there are any things, become intelligible by being first perceived, and then *abstracted* from our crude sense-perception. What we regard as the *real* world is not science, but it is that from which, by the aid of our minds and senses, we extract the accurate, reliable, and well-tested assertions which are the raw material of Science. Science itself consists of these assertions, linked into intelligible schemes which show how they are connected, and how new and reliable assertions flow from and are linked to them. In a sense natural science might link up every observation in one vast structure, but while Science is a-building, which as far as we can see will be for ever, it will remain divided into a number of not very closely linked departments.

As we have seen, the new science of the sixteenth century (Ch. VI) was little more than description, and there was, except in astronomy, very little explanation. Accordingly the new scientific men of the early seventeenth century (Ch. VII) had to find a scheme by which their observations could become *science*.

We suppose that there is only one real world from which we draw our science; but it does not follow that our science is the only science; for if that real world exists apart from the human observer, it doubtless has many aspects which we could but do not study, and probably many more of which we are totally unaware.

But, regarding science as a way of studying nature, as we know it, the two chief questions we may ask about a thing are:

(1) *What is it for?*
(2) *How and of what is it constructed and how does it change?*

Thus the flower of a lily may be regarded:

(1) As an arrangement for producing fertile seeds and per petuating the race.

(2) As a cone-shaped structure built of cells of such-and-such internal constitution and mode of operation.

Both of these are valid and valuable accounts of the flower; but when we come to study the phenomena of physics or chemistry, the conduction of heat or the rusting of iron—we do not to day find ourselves asking the purpose of these phenomena, but only how and when they occur.

The science of Aristotle and of the scholastics had been primarily a science of purpose. It was transparently clear to the latter that the world and all that's in it was created for the service of man, and that man had been created for the service of God. That was a perfectly intelligible scheme of the world. The sun was there to give us light and to tell the time and mark out the calendar by his motions. The stars and planets were a means of distributing beneficent or maleficent influences to the things on earth to which they were sympathetically linked: plants and animals were there to give us food and pleasure, and we were there to please God by doing his will. Modern science does not produce any evidence that all this is not perfectly true, but it does not regard it as a scientific explanation; it is not what science wants to know about the world.

The usual way of explaining the behaviour of things in the Middle Ages was to suppose they had been gifted with a "natural appetite" to do what they were known to do. Thus, "He who gave gravity to stones inclined them to fall naturally downwards." Yeast caused dough to rise because it had a natural fermentative virtue: poppies caused sleep through their inherent quality of coldness. The disadvantage of this kind of explanation was that it took you no further. The fermentative virtue, or the coldness, or the appetite to fall, were "occult qualities" and there was no way of examining them further; and the whole system amounted to no more than saying that the various objects in the world behaved according to a hidden nature implanted in them by God at the creation. This the succeeding generations would not deny, but they believed that the world when studied more closely would prove to be more orderly and reasoned, and less arbitrarily ordained than it had seemed to be.

The Mathematical World

A new way of looking at Nature was needed, but complete novelties are extremely rare events in the history of man. The revival of science, as we have seen, came from the new study of Greek texts, and of these the most admirable were the works of the Greek mathematicians. So it was very natural that when the reading of the works of Archimedes stimulated Galileo to re-found physics, he should follow the methods of his master and make his physics a branch of mathematics. Indeed, what we call a physicist was, at this period, termed a mathematician. Galileo was "First Mathematician to the Duke of Tuscany". So the ideal of the period was to express the foundations of physics (chiefly dynamics, statics and hydrostatics) as mathematical propositions; and so successful was this, that it seemed to the scientists of the time that all nature could be understood mathematically. All Galileo's physical discoveries, except that of the telescope, are concerned with measurements or numbers, and he is the first to present the scientific view of the world in the form of mathematical formulæ. This is a very fruitful practice, because all mathematical expressions are exactly and fully intelligible to all who take the trouble to learn mathematics, while the same cannot be said of any other mode of describing things. Descartes enthusiastically followed Galileo, and all the scientific men of the century who worked in any field capable of mathematical treatment resorted to mathematical methods (cf. Harvey, p. 102).

Now mathematical formulæ cannot deal with anything other than numerically measurable or geometrically specifiable data, such as size, shape, position, velocity, weight; and they can have no concern with the other things we perceive and infer in the real world, such as beauty, purpose, dignity and the like—all of which were very real and important qualities to the mediæval man of science and are still so to men and women in their daily lives. Thus the expression of science in terms that could be mathematically investigated meant the omission from science of all those elements of the real world that could not be quantitatively expressed.

Atomic Theories

It was obvious to all the seventeenth-century scientists that a number of common phenomena did not follow in any obvious way from the size, shape or movement of a body. Such were its degree of heat,* its colour, its lustre, its density, hardness, combustibility, solubility, fusibility, and so on.

Of these properties of bodies there were two standard explanations:

(1) Aristotle's theory that all bodies were composed of a continuous prime matter, modified into a composition of the four 'elements'—(earth, air, fire and water) and specified by various 'forms' so as to produce a material of the properties we observe.

(2) The theory of Democritus, that matter was discontinuous, being composed of many different kinds of atoms, whose various shapes and sizes and motions could account for all the properties of all kinds of matter.

Aristotle's theory prevailed in the Middle Ages, partly because it was a part of his system (adopted for other reasons), partly because the atomists of antiquity—Democritus, Epicurus, Lucretius—had been anti-religious, believing as they did that the chance meetings and associations of atoms determined all the happenings of the world, including those of the human mind. Moreover no one had ever seen these atoms and any phenomenon could be explained by inventing a new sort of atom to account for it, which made the theory a little unconvincing.

But in the atomic system the seventeenth century saw a way of explaining the whole world mathematically. If atoms were of known shapes, sizes and weights, and all phenomena were the result of atoms pressing or striking against other atoms, then every physical effect followed from its cause by necessity and in a way which ideally could be calculated from the laws of mechanics.

The atomic theory of the seventeenth century differed from the present one in many ways, including the following:

* Temperature had a rather different meaning in the sixteenth century.

(1) The atoms were regarded as hard, imperishable particles like so many minute shot or sand-grains.

(2) They had a precise shape, round, square, pointed, hooked, spiral, or the like.

(3) They were not believed to exert any field of force.

(4) They were not generally or necessarily thought to be in any rapid motion.

(5) Many, but not all, scientists considered atoms were permanently in contact; and that the world was completely full of them.

Galileo and Bacon both believed in atoms and the latter reasoned out that heat was a motion of atoms. Descartes in his *Principia Naturæ* tried to explain all the phenomena of the world in terms of the motions of the various kinds of atoms, and although his success was not marked, the book was very in-

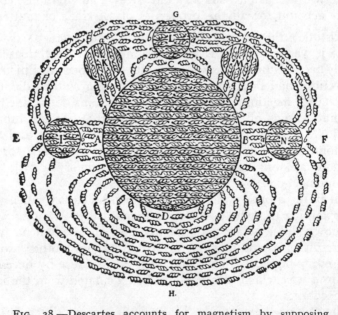

FIG. 38.—Descartes accounts for magnetism by supposing streams of screw-threaded particles passing through threaded passages in the earth and in iron, so aligning these as is seen in the effect of the earth on the compass needle.

fluential in making others study the matter more closely. Gassendi carried the matter further, and in this country Robert Boyle, who studied a great variety of chemical and physical phenomena, tried to explain them all in terms of atoms.

The Mechanical Philosophy

Boyle's view of the world (c. 1660) is very typical. He is convinced that it is the direct creation of God, but for him, unlike the mediæval philosophers, this is not enough; for, as he says, a man would be a dull fellow if, when he wanted an explanation of a watch, he was to be satisfied with being told it was an instrument made by a watch-maker—true though it is. An intelligent man, Boyle thinks, would want to know how the spring and balance and wheel and hands co-operate to form a time-telling instrument. So he gives us his opinion that "the world being once constituted by the great Author of things as it now is, *I look upon the phenomena of nature to be caused by the local motion of one part of matter hitting against another.*"

Here, then, is the "mechanical philosophy", as it came to be called. Everything is to be explained by the motion of the particles of bodies.

Newton followed Boyle in this view, but makes a further advance. He had shown (Ch. VII) how the assumption of universal gravitation, acting according to the inverse-square law, explained the motions of the planets in the skies; and so he took a further step towards our present view by assuming that atoms act as centres of force, so that an atom would not have actually to press against or touch another atom to put it in motion. The motions of atoms acted on by mutual forces could in theory be calculated, and so for Newton, every natural phenomenon, which could be the subject of science, whether on the earth or in the heavens, was ultimately to be explained by *mathematical laws*, deduced from reliable observation of nature and verifiable by further observation; ideally nothing else than accurate observation and mathematical reasoning was to be admitted into science, which would therefore consist of exact truth about the observable material world, though it would not be complete truth about the real world.

Thus according to Newton there was hope that every phenomenon, if knowledge were far enough advanced, "could be explained by the action of forces representing either attraction or repulsion depending only upon distance and acting between unchangeable particles."*

Actually this view, when applied in the limited field to which it belongs, was near enough to the truth to serve for a couple of centuries, but to-day it has had to be a good deal modified (Chs. XVII, XIX).

Science and the Mind of Man

Hardly anyone in the seventeenth century was rash enough to try to explain the human mind or soul in terms of the movements of atoms. Generally a very sharp distinction was drawn between the material body and the immaterial soul. Descartes regarded the body as a machine, the soul being by the special act of God in communication with and in control of it through a special organ, the pineal gland. Then as now, everyone was completely at a loss to explain how the immaterial human mind was related to the material body. In the Middle Ages, and later, the body was not regarded as a machine. The soul was the form of man, the body stood to it in the relation of matter, so there was no part of the body that was not informed by soul. But when Descartes and his followers put forward the theory that the body was a machine completely governed by the laws that governed phenomena outside it, he had to restrict the soul to a minute portion of it and to suppose that the soul received its information about the world through the brain and senses; and consequently it followed that this information might be greatly coloured or distorted by the perceiving and transmitting mechanism.

This fitted in very happily with the views of Galileo and others that were being expressed about this time. In science, as conceived by them, there was only room for the mathematically expressible properties of size, shape, and motion. But consider a peacock's feather. We perceive its size and shape, but no less clearly do we perceive its colour, lustre,

* Einstein and Linfield. *The Evolution of Physics.*

FIG. 39.—Descartes' diagram showing how the eyes perceive the arrow, and the soul in the pineal gland transmits an impulse to the muscles, causing the finger to point at the arrow. (From the 1677 edition of Descartes' *De Homine*, written before 1650.)

beauty. Galileo distinguished the qualities of the feather that did not depend on the human senses from those that did. Redness, greenness, lustre, beauty, cannot be thought of existing apart from an eye and a brain, but size, shape and motion could. So the latter he regarded as the real or primary qualities to be studied by science and the former as little more than illusions or at most properties of one little corner of the world—the human brain.

This view has remained substantially the view of science till the present day. The objects of the scientists' 'real world' have size, shape, motion, but there is no colour, scent, beauty in that world: there is no purpose, no emotion, no love. That is why it is so utterly dangerous to treat the world which the scientist has abstracted for study from what he perceives, as if it were the totality of things; and that is the fallacy of the materialist.

Because science is so intelligible, works so perfectly, and gives us such power, we are tempted to treat its world as the real world, despite our knowledge that the things which are important in human life are not quantitative or extended and so can never enter into science.

Examples Illustrating the Seventeenth-Century View of the World

GALILEO REGARDS THE UNIVERSE AS MATHEMATICAL

Philosophy* is written in that very great book, which continually lies open before our eyes (I mean the Universe); but we cannot understand it, if we do not first learn the language, and comprehend the characters in which it is written. It is written in the mathematical language, and its characters are triangles, circles and other geometrical figures, without the aid of which it is impossible to understand a word of it, without which one wanders vainly through a dark labyrinth.
(*Il Saggiatore*, Galileo Galilei, p. 171 of Florentine Edition.
1842.)

GALILEO DISTINGUISHES THE HEAT WHICH WE FEEL FROM THAT WHICH SCIENCE STUDIES

But first of all I desire to consider the thing that we call heat, concerning which I doubt very greatly if the generally accepted concept be not very far from the truth, inasmuch as it is believed to be a true accident, affection, and quality which really resides in the matter which we perceive to be heated. However, I say that I feel myself necessarily compelled, as soon as I conceive of a piece of matter or a corporeal substance, to conceive of it as determined and shaped as of such and such a figure, as being great or small in relation to other bodies, as being in this or that place, at this or that time, as being in motion or at rest, as touching or not touching another body,

* i.e. Natural Philosophy, which is the seventeenth-century term for what we call science.

as being one or many; nor can it by any imagination be separated from these conditions: but that it must be white or red, bitter or sweet, sounding or mute, of pleasant or unpleasant smell, I do not find myself mentally forced to consider it as necessarily accompanied by such conditions: so if the senses were not its escorts, perhaps neither reason nor imagination by itself would have arrived at them. Thence I am led to think, that these tastes, odours, colours, etc., from the point of view of the objects in which they appear to reside, are no more than pure names, and have their only residence in the sensitive body, so that if the animal* were removed all such qualities would be removed and annihilated. . . . (Therefore) I say I am much inclined to believe that heat is of this nature, and that the thing which in us produces and makes us feel heat and to which we give the general name of Fire, is a multitude of very small corpuscles figured in such and such a manner, moved with such and such velocity which encounter our body and penetrate it by reason of their exceeding fineness; and that the contact made by their passage through our substance and felt by us, is the affection that we call heat. . . . But that besides their form, number, movement, penetration and contact, there is in fire another quality, that is heat, I do not believe . . . and I judge that it is so far due to ourselves that if the animated and sensitive body were taken away, 'heat' would remain as no more than a mere word.

(*Il Saggiatore.* Galileo Galilei. Florentine Edition. 1842. pp. 333–336.)

(This view of heat is not the modern one. Galileo considers it to be a subtle material substance; we hold it to be the energy of motion of the particles of ordinary matter. None the less, the distinction between this motion and our sensation remains a true one.)

BOYLE ACCOUNTS FOR THE QUALITIES OF THINGS
BY AN ATOMIC HYPOTHESIS

We teach then (but without peremptorily asserting it):

* i.e. the observer.

1. That the matter of all natural bodies is the same; namely, a substance extended and impenetrable.

2. That, all bodies thus agreeing in the same common matter, their distinction is to be taken from those accidents* that do diversify it.

3. That motion, not belonging to the essence of matter (which retains its whole nature when it is at rest) and not being originally producible by other accidents, as they are from it, may be looked upon as the first and chief mood or affection of matter.

4. That motion, variously determined, doth naturally divide the matter it belongs to into actual fragments or parts; and this division, obvious experience (and more eminently chymical operations) manifest to have been made into parts exceedingly minute, and very often too minute to be singly perceivable by our senses.

5. Whence it must necessarily follow, that each of these minute parts or *minima naturalia* (as well as every particular body made up by the coalition of any number of them) must have its determinate bigness or size, and its own shape. And these three, namely, bulk, figure, and either motion or rest (there being no mean between these two) are the three primary and most catholick moods or affections of the insensible parts of matter, considered each of them apart.†

6. That when divers of them are considered together, there will necessarily follow here below both a certain position of posture in reference to the horizon (as erected, inclining or level) of each of them, and a certain order or placing before or behind or beside one another . . . and when many of these small parts are brought to convene into one body from their primary affections, and their disposition and contrivance as to posture and order, there results that which by one comprehensive name we call the texture of that body. . . . And these are the affections that belong to a body, as it is considered in

---

* Accidents: properties not essential to the substance, as coldness to water or crystalline form to sulphur. Here the form and motion of the atoms is intended.

† i.e. considering only one such minute part of matter.

itself, without relation to sensitive beings or to other animal bodies.

7. That yet there being men in the world whose organs of sense are contrived in such differing ways, that one sensory* is fitted to receive impressions from some, and another from other sorts of external objects or bodies without them (whether these act as entire bodies, or by emission of their corpuscles, or by propagating some motion to the sensory) the perception of these impressions are by men called by several names, as heat, colour, sound, odour; and are commonly imagined to proceed from certain distinct and peculiar qualities in the external objects, which have some resemblance to the ideas their action upon the senses excites in the mind; though indeed all these sensible qualities, and the rest that are to be met with in the bodies without us, are but the effects of consequents of the above-mentioned primary affections of matter, whose operations are diversified according to the nature of the sensories or other bodies they work upon.

8. That when a portion of matter, either by the accession or recess of corpuscles, or by the transposition of those it consisted of before, or by any two, or all of these ways, happens to obtain a concurrence of all those qualities which men commonly agree to be necessary and sufficient to denominate the body which hath them, either a metal or a stone, or the like, and to rank it in any particular and determinate species of bodies, then a body of that denomination is said to be generated.

(*The Origin of Forms and Qualities according to the Corpuscular Philosophy*. Robert Boyle, 1667.)

(This passage should be carefully compared with our modern atomic·theory. Note the absence of an idea of a field of force surrounding the atoms (present in Newton's account, below). It is to be remembered that there was as yet no theoretical chemistry nor were any particular substances characterised as chemical elements, in the modern sense.)

---

* The part of the brain which receives the reports of the sense-organ and by which they are consciously perceived.

NEWTON ON ATOMS AND OCCULT QUALITIES

All these things being consider'd, it seems probable to me, that God in the Beginning, form'd Matter in solid, massy, hard, impenetrable, moveable Particles, of such Sizes and Figures, and with such other Properties, and in such Proportion to Space, as most conduced to the End for which he form'd them; and that these primitive particles being Solids, are incomparably harder than any porous Bodies compounded of them; even so very hard as never to wear or break in pieces. No ordinary power being able to divide what God himself made one in the first Creation. While the Particles continue entire, they may compose Bodies of one and the same Nature and Texture in all Ages: But should they wear away, or break in pieces, the Nature of Things depending on them would be changed. Water and Earth composed of old worn Particles and Fragments of Particles, would not be of the same Nature and Texture now, with Water and Earth composed of entire Particles in the Beginning. And therefore that Nature may be lasting, the Changes of Corporeal Things are to be placed only in the various Separations and new Associations and Motions of these permanent Particles: compound bodies being apt to break, not in the midst of solid Particles, but there those Particles are laid together, and only touch in a few Points.

It seems to me farther, that these Particles have not only a *Vis inertiæ* (inertia), accompanied with such passive Laws of Motion, as naturally result from that Force, but also that they are moved by certain active Principles, such as is that of Gravity, and that which causes Fermentation, and the Cohesion of Bodies. These Principles I consider not as occult Qualities, supposed to result from the specifick Forms of Things, but as general Laws of Nature by which the Things themselves are formed; their Truth appearing to us by Phænomena, though their Causes be not yet discovered. For these are manifest qualities and their Causes only are occult. And the *Aristotelians* gave the Name of occult Qualities Not to manifest Qualities, but to such Qualities only as they supposed to lie hid in Bodies, and to be the unknown causes of manifest Effects: Such as would be the Causes of Gravity, and of magnetick and electrick Attractions, and of Fermentations, if we should suppose that the forces or Actions arose from Qualities unknown to us and uncapable of being discovered and made manifest. Such occult qualities put a stop to the Improvement of natural Philosophy, and therefore of late Years have been rejected. To tell us that every species of Thing is endow'd with an occult specifick Quality, by which it acts and produces manifest Effects, is to tell us nothing: But to derive two or three general Principles of Motion from Phænomena, and afterwards to tell us how the Properties and Actions of all corporeal Things follow from these Principles would be a very great step in Philosophy, though the Causes of those Principles were not yet discovered: And therefore I scruple not to propose the Principles of Motion above-mentioned, they being of very general Extent, and leave their Causes to be found out.

Now by the help of these Principles, all material Things seem to have been composed of the hard and solid Particles above-mentioned, variously associated in the first Creation by the Councel of an intelligent Agent. For it became him who created them to set them in order. And if he did so, it's unphilosophical to seek for any other origin of the World, or to pretend that it might arise out of a Chaos by the mere Laws of Nature; though being once form'd, it may continue by those Laws for many Ages.

(*Opticks*: or, *A Treatise of the Reflections, Refractions, Inflections and Colours of Light*. 1704. Sir Isaac Newton. pp. 375–8.)

NEWTON SUSPECTS THAT ATOMS EXERT FORCES

We offer this work as mathematical principles of philosophy . . . by the propositions mathematically demonstrated in the first book we then derive from the celestial phenomena the forces of gravity with which bodies tend to the sun and the several planets. Then, from these forces, by other propositions which are also mathematical we deduce the motions of the planets, the comets, the moon and the sea. I wish we could derive the rest of the phenomena of nature by the same kind of reasoning from mechanical principles; for I am induced by many reasons to suspect that they all depend upon certain forces by which the particles of bodies by some causes hitherto unknown, are either mutually impelled towards each other and cohere in regular figures, or are repelled and recede from each other; which forces being unknown, philosophers have hitherto attempted the search of nature in vain. But I hope the principles here laid down will afford some light either to that or some true method of philosophy.

(*Principia Mathematica.* Sir Isaac Newton. Tr. La Motte, Part I.)

CHAPTER NINE

The First Principles of the Sciences

The development of laboratory technique

Even before the year 1700 everyone recognised the scientific method as the only appropriate way of investigating the external world. But the scientific method gives no more than an outline of the way to plan and conduct a scientific research, and is of no help towards the devising of the practical laboratory methods by which discoveries are made. The scientists of the eighteenth century had therefore to occupy themselves in discovering practical laboratory techniques, and for this reason the period often seems a little disappointing in its results, as compared with the seventeenth and nineteenth centuries.

The most notable advance in scientific technique was the discovery of numerous ways of recording the behaviour of things in terms of *numbers*. Measurement was, of course, ancient enough. The standard yard, bushel and gallon went back to at least the thirteenth century, but the official standards of length, capacity and weight, were intended rather for commerce than for science, and it was not until 1747 that the Royal Society was called in by the government to construct standards. This became the more necessary when measuring instruments of many kinds were invented. In the period 1650–1700 instrument-makers capable of fine and accurate craftsmanship were to be found in every great town. Thus accurate measurement of *length* was made possible by the vernier scale (1631) and micrometer (1638). These discoveries also permitted precise measurements of angles. The accurate measurement of *time* was an enormous advance. An exact standard of time was always available, namely the rotation of the earth, but the only means of discovering the time before the middle of the seventeenth century was an astronomical observation. Well-designed apparatus could do this very closely, but portable apparatus, e.g. astrolabes, quadrants and sundials, had an error of at least five minutes. Clocks before c. 1660 had errors of ± 20 minutes a day, but when Huygens applied the pendulum, and Hooke the balance-spring, to clocks, their errors were reduced to those we find in the ordinary domestic clocks of to day. But the effect of changes in temperature upon the length of pendulums and the stiffness of springs prevented really accurate time-keeping until Harrison discovered methods of compensation and gave us chronometers that kept time to a minute or so a month (1757). *Weights* had always been pretty accurately measurable in jewellers' balances, but these were also greatly improved. The pendulum clock revealed the fundamental distinction of mass and weight. When Jean Richer took his astronomical apparatus, including a pendulum clock, from Paris to Cayenne in 1671 he found the clock lost $2\frac{1}{2}$ minutes a day. The clock was corrected, but when brought back to Paris it gained $2\frac{1}{2}$ minutes a day. The gravitational force that attracted the pendulum-bob was different at Cayenne, but its inertia was the same. The effect was explained by Huyghens, whose work *On the Vibrating Pendulum* is second only to Newton's *Principia*.

Length, time and mass are the fundamental units, but instruments for measuring other quantities were soon available. Thus alcohol-thermometers, standardised in accordance with arbitrary standard thermometers, were available from the middle of the seventeenth century, though the use of natural fixed standards, such as the temperature of melting ice or boiling water, dates only from soon after 1700. The mercury-barometer was in use from about 1645, and pressure-gauges soon followed. Photometry, the measurement of quantity of light, was also a seventeenth-century invention, and simple electrometers were also in use. Thus the seventeenth and eighteenth centuries provided *the means of obtaining and recording numerical results*.

Scientific Explanations in the Eighteenth Century

As we have seen, the theory that matter is made up of atoms endowed with attractive and repulsive forces was suggested by Newton in his *Principia*; but, as he recognised, science was then very far from being able to explain the phenomena it studied in terms of these atoms and the forces acting on them. Heat, light, electrical and magnetic phenomena, gravity, chemical affinity, muscular action: all these seemed to be disconnected separate principles, and although the theory that matter consisted simply of atoms in motion and exerting forces could be believed, it was not yet a practically useful guide to laboratory experiments. But all the above are *fundamental principles*; and it is difficult (though not impossible) for example, to study heat without some working hypothesis as to what makes the difference between a hot body and a cold one. Even today we have not elucidated the whole of these problems, but we can classify and connect all the above phenomena as being manifestations of *energy*, the capacity for doing work. The eighteenth century, on the other hand, regarded heat, electricity, magnetism, and chemical energy as *imponderables*—material fluids, in some ways like a gas or liquid, but invisible, without weight, and capable of flowing into and permeating solids or liquids: thus Lavoisier places heat and light in his tentative list of chemical elements.

The fact that these fluids do not really exist does not render eighteenth-century science worthless; for most of the men of science of that age realised that the evidence for the presence of "caloric" in hot bodies or "electric fluid" in electrified ones was, to say the least of it, scanty, and they were consequently more concerned to observe and record how things behaved than to explain their behaviour. So from simple laboratory experiments the seventeenth and eighteenth centuries learnt a great deal of correct fact, of which the nineteenth century gave correct explanations.

Progress in the individual Sciences

(a) *Mechanics.* The fundamental principles had been set out by Newton, and although there was little engineering to stimulate it until nearly the end of the eighteenth century, statics, hydrostatics, and dynamics advanced very rapidly in the hands of the numerous brilliant mathematicians of the age. The other physical sciences lagged far behind because their fundamental principles were not understood and the results of experiments often seemed disconnected and irrational.

(*b*) *The study of heat.* Thus in the study of heat very little could be done until the notions of quantity of heat and temperature were distinguished. 'Heat' remained a confused notion, not properly separated from what we now call flame, combustion, chemical energy, until after the turn of the century. But this did not stop the experimentalists from measuring the effects of temperature changes. Coefficients of expansion were carefully measured; these were of practical importance for compensating chronometer-springs, pendulums, surveying-chains, etc., for errors resulting from temperature changes. Joseph Black did work of fundamental importance when he distinguished quantity of heat and temperature, and gave us the notions of specific and latent heat. Lavoisier and Laplace followed him and founded calorimetry. Radiant heat was studied: the existence of it was a difficulty for the caloric theory (p. 158), and by 1800 it seemed that heat-rays might be a kind of invisible light. It was clearly proved that heat had no weight and was produced by friction, compression of gases, etc.: and when Rumford in 1798 was able to produce an apparently unlimited amount of heat from friction (pp. 159–60) it seemed difficult to retain the caloric theory, which none the less was not yet extinct in 1850.

The application of this to the practical use of heat in the steam-engine will be considered in Chapter XI.

(*c*) *The study of light.* The facts of the transmission of light were also carefully studied, and the geometry of the transmission of light, its refraction, reflection, dispersion, etc., were worked out. The great practical advance was in the design of optical instruments. Single lenses necessarily give colour-fringed images, which render impossible the use of high powers in microscopes and telescopes. The reflecting telescope dispensed with lenses in favour of mirrors, and this instrument, invented by Newton, was brought to a high degree of perfection in the eighteenth century. In 1758 Dollond discovered a way of avoiding these colour-fringes, by making achromatic lenses compounded of two kinds of glass. These soon came into general use in telescopes, but not in microscopes until c. 1825. These lenses gave much sharper images; this enabled microscopists to use higher powers. The use of improved microscopes led in the early nineteenth century to great discoveries in biology, notably the cell-theory; while the improved telescopes led to the study of nebulæ, the discovery of the asteroids and of the two great outer planets. All this was the result of experimental work, and was independent of the question as to whether light consisted of waves or particles.

(d) *The study of electricity.* Electricity excited great interest. At the beginning of the eighteenth century frictional electric machines were invented. These gave extremely minute currents at very high voltages, but even so they allowed of the discovery of some of the main phenomena, such as electrical attraction and repulsion, the distinction between conductors and non-conductors, the use of condensers such as the Leyden jar, and the discovery of the electric shock. Benjamin Franklin made the great discovery that lightning was no more than a huge electric spark, and so invented the lightning-conductor. Coulomb, from 1784 on, began the quantitative study of electricity, *measuring* the repulsion between charges and proving that this followed the inverse-square law. Little further could be done till 1800 when Volta invented the electric battery, for only then could electrolysis and electromagnetism be studied.

(e) *The study of chemistry.* Chemistry was in the same plight. Active experimentalists, such as Scheele and Priestley, discovered new and important compounds by the dozen, and the ideas of the great classes of acids, bases, salts, earths, etc., became well defined and their combining proportions began to be studied. A beginning was made in chemical analysis. Nevertheless, there was no clear notion of the relation of one compound to another. The idea of chemical elements and compounds had been initiated by Boyle in the seventeenth century, but until the time of Lavoisier nobody had tried to make a list of supposed elements. The Aristotelian four elements and the alchemical three principles (salt, sulphur and mercury) were discredited, but there was nothing to put in their place.

Little progress in chemical theory was made, because the all-important element, oxygen, was unknown, and a hypothetical 'matter of fire', *phlogiston*, was invented to account for the phenomena of oxidation and reduction. Thus the rusting of iron or burning of charcoal really involve a *combination* with oxygen; but they were explained as the *loss* of this hypothetical phlogiston. Worst of all, the most universal and important of elements—oxygen—remained unknown until 1774. After Lavoisier had realised the importance of this element, chemistry could be explained in terms of the combination and separation of known material elements, and it is here that modern chemistry begins. These matters will be taken up again in Chapter X.

(*f*) *The study of biology.* We see much the same in the study of living things. Their classification and the mapping of their anatomy went ahead rapidly, but their physiology, i.e. how they worked, remained nearly stationary. The fundamental processes of animals and plants are chemical reactions taking place in their individual cells, and until there was a little elementary chemical theory and microscopes good enough to demonstrate more than the most obvious of cell-structures, not much could be done.

Need for the conception of Energy

In each of these sciences there was a similar difficulty, which we may put down to the lack of the notion of *energy*. Kinetic energy, *vis viva*, was known to the men of the eighteenth century, but they had no idea that heat, light, magnetism, electricity, and the power that lay hid in coal or gunpowder, were all manifestations of this same energy, which was as *real* as matter. What is the difference between a hot body and a cold one, between an electrified piece of sealing-wax and an 'unexcited' one, between a magnet and an ordinary piece of steel? We say that it is some movement or disposition on the parts of the bodies—molecules, atoms, electrons—which give the hot, the magnetised, the electrified, more *energy* than the cold, unmagnetised, or non-electrified. The eighteenth-century answer was very different. It was supposed that the active forms—the hot, the electrified, the magnetised—contained a substance *material, but subtle and weightless*. A hot body contained the 'matter of heat' or the *caloric fluid*; electrified wax contained an excess or deficiency of the *electric fluid*; a magnet the *magnetic fluid*; coal or gunpowder contained *phlogiston*; a living organism contained *vital spirits*, while a dead one did not.

All these subtle fluids or spirits were thought of as material, weightless, more 'subtle' than a gas, and so capable of penetrating into bodies without any visible apertures. They were thought to form a material link by which two bodies that attracted or repelled each other were drawn together or driven apart. Thus it seemed intelligible that two bodies might repel each other by means of subtle fluids which passed out of them and forced them apart, but no one at the time could see how two bodies with nothing in between them could affect one another. Nor can we: we are completely in the dark as to the process or mechanism by which the magnet makes the iron filings move. But we do not mind taking for granted phenomena for which we cannot picture or suggest any mechanism. The eighteenth century, the age of reason, had to invent these fluids in order to give a rational picture of what was happening.

The theory worked pretty well for electricity, because it does actually consist of electrons which do move about conductors, very much as a fluid might move. For heat it was inadequate because it could not explain the connection between heat and work, nor were the other 'imponderables' of much more use in their respective sciences.

They served their turn, however, in providing a means by which men could picture these energy-phenomena, and as a means of prompting further experiments. When truer notions of heat, light, electricity, chemical energy came into being, the *facts* discovered and explained by the aid of the older ideas still remained true, and the terminology was easily changed. No matter what happens to scientific theory, scientific fact remains; and in this is the substance of science. A part of the great structure may need to be rebuilt in a new style from time to time, but the hewn stone of fact has simply to be rearranged.

What was the fate of these imponderables? The notion of

the caloric fluid died hard, but when it became clear that heat was converted quantitatively into work and *vice versa*, it became quite untenable, because work was certainly not a material substance at all: yet for all that the caloric theory was not quite extinct in the eighteen-fifties. The 'matter of light' was no more heard of after the eighteen-thirties, when the wave theory, which since the time of Huygens had always had its exponents, was generally accepted. The notion of phlogiston and of the vital spirits died out after 1790 when Lavoisier had explained the essentials of respiration and combustion. The electric fluid lasted longest. Electricity remained a great puzzle through most of the nineteenth century, for it seemed to behave like a real fluid and also like energy. The theory that it consisted of electrons—particles with great mutual repulsive forces —only came into general use after 1897. We are to-day still left with the puzzle that there are three independent kinds of attraction, electric, magnetic, and gravitational, and though Einstein may have incorporated them into a single scheme of description, we have as yet no notion as to how any of them operate.

Conservation of Mass and Energy

A very great step was the enunciation of two great principles, the Conservation of Matter and the Conservation of Energy.

The idea that matter is not created or destroyed in any reaction, but only changed, had been long suspected, and indeed acted on in chemical practice. Newton's views on atoms necessarily imply it (p. 142). The law cannot be *proved*, for this would involve proving a universal negative. The evidence for it is the absence of any evidence for creation or destruction of matter in any of the countless chemical experiments in which it might have been detected, if it had in fact occurred. Lavoisier rightly assumes it as an *axiom* " . . . nothing is created in the operations either of art or of nature, and it can be taken as an axiom that in every operation an equal quantity of matter exists both before and after the operation . . ."

The law of the Conservation of Energy was a much more sweeping generalisation. The notion of kinetic energy had been developed by Huygens in the seventeenth century, and the law that the quantity of mechanical energy in a system remained unchanged was employed in eighteenth-century mechanics. After the end of the eighteenth century the "convertibility of the powers of nature" began to be appreciated. Chemical energy could be changed into electrical in the voltaic battery; electricity into heat and *vice versa*; magnetism into electricity and *vice versa* by the interaction of magnets and conductors; heat into mechanical energy in the steam-engine; heat into light and *vice versa*. How then was it possible to regard heat, light, electricity, etc., as different fluids? Quantitative studies settled the question. Joule showed a numerical connection between quantity of electricity and quantity of heat produced by it, and between the quantity of work and the quantity of heat produced by it. The final establishment of the principle came from Helmholtz as late as 1847. He accepted the above-mentioned transformations of the powers of nature one with another. He assumed that it was impossible, as experience also showed, to obtain an indefinite amount of work from any closed system of bodies: he also assumed that all the activities of nature could be derived from attractive and repulsive forces acting between material particles. From such assumptions he showed that *the sum of the kinetic and potential energies in any such system must always be the same* and consequently that energy was never lost, but only transformed from one state of manifestation to another.

Science in the eighteenth century

The eighteenth century may seem to us a disappointing period. The seventeenth century seems, and indeed was, a brilliant age of science, but it is to be remembered that its triumphs are to be contrasted with earlier ages when the scientific method did not exist. Startling discoveries and grand generalisations were made, but there was not enough talent, time, or opportunity, to make more than a beginning. Thus the first three-quarters of the eighteenth century were taken up with observations and experiments, which provided a foundation for the individual sciences: this unspectacular work was the basis of the great harvest of discovery which filled its closing years. Throughout the century there was little professional science teaching and few whole-time men of science, except in the medical profession; the work was mostly carried on by enthusiastic amateurs, many of them clergymen.

Science began to be popularised in the eighteenth century, and numerous books, encyclopædias, etc., were written, not for the learned, but for the general public. The chief matter of public interest was the Newtonian system: as always, people wanted to know what sort of a universe they inhabited. From these accounts some drew the conception of a universe made by a good and orderly Creator, while others were led to regard the All as a vast inanimate machine. The former school was most conspicuous in England and Germany, the latter in France.

Examples of Eighteenth-Century Science

ELECTRICITY AND LIGHTNING. A LETTER OF BENJAMIN FRANKLIN, ESQ., TO MR. PETER COLLINSON, F.R.S., CONCERNING AN ELECTRICAL KITE

Philadelphia, Oct. 1st, 1752.

As frequent mention is made in the public papers from Europe of the success of the Philadelphia experiment for drawing the electric fire from clouds by means of pointed rods of iron erected on high buildings, etc., it may be agreeable to the curious to be informed that the same experiment has succeeded in Philadelphia, tho' made in a different and more easy manner. which any one may try, as follows:

Make a small cross of two light strips of cedar; the arms so long, as to reach to the four corners of a large thin silk handkerchief, when extendeed: tie the corners of the handkerchief to the extremities of the cross; so you have the body of a kite; which being properly accommodated with a tail, loop and string, will rise in the air like those made of paper; but this, being of silk, is fitter to bear the wet and wind of a thunder gust without tearing.

To the tip of the upright stick of the cross is to be fixed a very sharp-pointed wire, rising a foot or more above the wood.

To the end of the twine next the hand, is to be tied a silk riband; and where the twine and silk join, a key may be fasten'd.

The kite is to be raised, when a thunder-gust appears to be coming on (which is very frequent in this country) and the person, who holds the string, must stand within a door, or window, or under some cover, so that the silk riband may not be wet; and care must be taken that the twine does not touch the frame of the door or window.

As soon as any of the thunder-clouds come over the kite, the pointed wire will draw the electric fire from them; and the kite, with all the twine, will be electrified; and the loose filaments of the twine will stand out every way, and be attracted by an approaching finger.

When the rain has wet the kite and twine, so that it can conduct the electric fire freely, you will find it stream out plentifully from the key on the approach of your knuckle.

At this key the phial* may be charged; and from electric fire thus obtain'd spirits may be kindled, and all the other electrical experiments may be performed, which are usually done by the help of a rubbed glass globe or tube, and thereby the sameness of the electric matter with that of lightning completely demonstrated.

I was pleased to hear of the success of my experiments in France and that they there begin to erect points upon their buildings. We had before placed them upon our academy and state-house spires.

(*Philosophical Transactions.* Vol. 47, p. 565.)

PRIESTLEY'S VIEW OF THIS DISCOVERY

There is nothing in the history of philosophy more striking than the rapid progress of electricity. Nothing ever appeared more trifling than the first effects which were observed of this agent in nature, as the attraction and repulsion of straws and other light substances. It excited more attention by the flashes of light which it exhibited. We were more seriously alarmed at the electrical shock, and the effects of the electrical battery;†

* Leyden jar.

† He refers to a battery of Leyden jars, not a voltaic battery.

and we were astonished to the highest degree by the discovery of the similarity of electricity with lightning, and the aurora borealis, with the connection it seems to have with water-spouts, hurricanes, and earthquakes, and also with the part that is probably assigned to it in the system of vegetation, and other the most important processes in nature.

(*Experiments and Observations on different kinds of Air.*
Joseph Priestley. 1774. p. 274.)

ACCOUNTS OF SOME OF THE IMPONDERABLES

The human mind, never satisfied, after the cause of some effect has been discovered, or only guessed at, attempts to investigate some more intimate quality, and even the origin of the supposed cause, making further suppositions, and framing other hypotheses, which, by the course of things, must certainly be less probable than the former. This unlimited endeavour to acquire knowledge is often too ridiculous to be pursued, on account of its abstruseness and uncertainty, especially when the steps immediately preceding the subject in hand have but a small degree of probability. It is from hence that philosophers have frequently spent a great deal of time, and trouble, in attempting to discover the properties and causes of what existed only in their own imaginations. Sometimes, however, when a supposed existence comes so very near to truth, that the most sceptic Philosopher hesitates not to confess the probability of it, or when he can invent no argument to evince the contrary, then it is not only allowable, but necessary for the business of Philosophy, to pursue the enquiry further, and, if nothing else can be ascertained, at least to propose some further conjectures upon the former hypothesis. This now is the case in the science of Electricity; and after we have related the most plausible hypothesis as yet offered, i.e. that of a single elastic fluid, we come in this place to consider the essence of this fluid, in order, if possible, that we might attain to, at least, some probable conjecture respecting its materials.

When nothing more than electric attraction and repulsion has been observed, Electricians supposed that these were effected by a kind of unctuous effluvia proceeding immediately from the electrified body; but when the light, the burning quality, the phosphoreal smell,* etc., was perceived to be produced by excited Electrics, then it was naturally supposed, that the electric fluid was of the same nature with fire. This opinion has prevailed much among several Philosophers, and it is from hence, that the electric fluid has been commonly called Electric Fire. Besides this supposed identity of the electric fluid, and the element of fire, there have been two other opinions concerning the essence of this fluid; it having been thought by some to be the *ether* of Sir Isaac Newton, and by others (whose opinion seems to be the most probable) to be a fluid *sui generis*, i.e. different from all other known fluids.

In order the more regularly to examine these conjectures, it will be necessary to premise something in regard to the nature of fire, at least so much as is sufficient for the present purpose.

The element of fire may be considered in regard to its spring, to the different states of its existence, and to its effects. In regard to its origin it is commonly specified under the names of Celestial, Subterraneous, and Culinary Fire; understanding by the first, that which proceeds from the sun, and, by being dispersed throughout the universe, gives life and motion to almost everything that exists; by the second, that which is the cause of volcanos, hot springs, etc., and lastly, under the name of Culinary Fire, understanding that which is commonly produced upon the earth by burning several substances. These distinctions, however, are little if at all useful; for, whatever be the origin of fire, its effects are always the same.

In respect to the different states of its existence, the Chymists know only two; the first obvious one, and indeed that, to which only is given the name of Fire, is that actual agitation of the particles of that element, which produces the complex idea of lucid, hot, etc. that is commonly understood under the name *Fire*; and the other state is the real principle of fire existing as a constituent principle in several, and perhaps all substances; or, that matter, whose particles, when agitated in a peculiar and violent manner, produce the common sensible fire.

* The odour of ozone, not yet identified.

This, which we may call fire in an inactive state, is the *Phlogiston* of the Chymists, and is that, which when united in a sufficient quantity with other substances, renders them inflammable. This this principle does really exist, is beyond a doubt; we may transfer it from one body to another; we may render a body inflammable, which in its own nature is not so, by superinducing on it the phlogiston; and we may reduce a body naturally inflammable, to a substance not inflammable, by depriving it of its phlogiston.

As to the identity of the Electric, and the ethereal fluid it seems to me quite an improbable, or rather a futile and insignificant hypothesis; for this ether is not a real, existing, but merely an *hypothetical fluid*, supposed by different Philosophers to be imbued with different properties, and to be an element of several principles. Some suppose it to be the element of fire itself, others make it the cause of attraction, others again derive animal spirits from it, etc.; but the truth is, that not only the essence, or properties, of this fluid but even the reality of its existence is absolutely unknown. According to Sir Isaac Newton's supposition, this ether is an exceedingly subtle and elastic fluid, dispersed throughout all the universe, and whose particles repel the particles of other matter. But on this supposition the electric fluid is different from ether; for, although the former is subtle, and elastic, like the latter, yet (as Dr. Priestley observes) it is not repulsive like the ether, but attractive of all other matter.

(*Treatise on Electricity.* Tiberius Cavallo. 1786.)

THE CALORIC FLUID

What we call *heat* is a sensation produced by a substance to which modern chemists have given the appellation of *caloric*. When caloric is applied to our system in a greater proportion than it already contains, the system is warmed, and the sensation of heat produced. When, on the contrary, a substance of a lower temperature than our system is applied to it, we feel the sensation of cold, because we then lose caloric.

Caloric penetrates all bodies; it separates their particles by lodging between them, and diminishes their attraction; it dilates bodies, it liquefies solids, and rarifies liquids to such a degree, as to render them invisible, give them the form of air, and convert them into elastic, compressible, aeriform fluids. Hence it follows, that liquids are combinations of solids with caloric, and that gases are solutions of different bodies in caloric, which of itself is the most attenuate, subtile, light, and elastic of all natural substances; accordingly its weight cannot be estimated. . . .

All these facts prove, that caloric is a particular substance, and not a modification of all substances, as some natural philosophers have imagined; and it is far from having been shown to be the same thing with light; for the farther we advance in the science of physics, the greater differences appear in the action of these two substances.

(*The Philosophy of Chemistry.* A. F. Fourcroy. 1800.)

RUMFORD IDENTIFIES HEAT AND MOTION

An Inquiry concerning the Source of the Heat which is excited by Friction. By Benjamin Count of Rumford, F.R.S., M.R.I.A.

. . . Being engaged, lately, in superintending the boring of cannon in the workshops of the military arsenal at Munich, I was struck with the very considerable degree of heat which a brass gun acquires, in a short time, in being bored; and with the still more intense heat (much greater than that of boiling water, as I found by experiment), of the metallic chips separated from it by the borer. . . .

From *whence comes* the heat actually produced in the mechanical operation above mentioned?

Is it furnished by the metallic chips which are separated by the borer from the solid mass of metal?

If this were the case, then, according to the modern doctrines of latent heat, and of caloric, the *capacity for heat* of the parts of the metal, so reduced to chips, ought not only to be changed, but the change undergone by them should be sufficiently great to account for *all* the heat produced—for taking equal quantities, by weight, of these chips, and of thin slips of the same block of metal separated by the saw, and putting them at the same temperature, (that of boiling water), into equal quantities of cold water (that is to say at the temperature of $59\frac{1}{2}°$ F.) the portion of water into which the chips were put was not, to all appearance, heated less or more than the other portion into which the slips of metal were put. . . . From this it is evident that the heat produced could not possibly have been furnished at the expense of the latent heat of the metallic chips.

(Rumford then designed an apparatus in which a blunt borer was forced against the bottom of a hollow metal cylinder and was rotated by machinery operated by horses. The cylinder was enclosed in a box containing about two gallons of water.) . . . At the end of an hour I found by plunging a thermometer into the water in the box that its temperature had been raised no less than 47 degrees: being now 107° of Fahrenheit's Scale. . . . At 2 hours 20 minutes it was at 200°; and at 2 hours 30 minutes it ACTUALLY BOILED!

It would be difficult to describe the surprise expressed in the countenances of the bystanders on seeing so large a quantity of water heated and actually made to boil without any fire.

Though there was, in fact, nothing that could justly be considered as surprising in this event, yet I acknowledge fairly that it afforded me a degree of childish pleasure, which, were I ambitious of the reputation of a *grave philosopher*, I ought most certainly rather to hide than to discover. . . .

And in reasoning on this subject we must not forget that the most remarkable circumstance, that the source of the heat generated by friction in these experiments, appears evidently to be *inexhaustible*.

It is hardly necessary to add, that any thing which any *insulated* body, or system of bodies, can continue to furnish, *without limitation*, cannot possibly be a *material substance*: and it appeared to me to be extremely difficult, if not quite impossible, to form any distinct idea of any thing, capable of being excited, and communicated, in the manner the heat was excited and communicated in the experiments, except it be MOTION.

(*Philosophical Transactions*. Vol. 88. 1798.)

CHAPTER TEN

The Study of Gases

The Importance of Gases

The beginner in science, to day, finds himself much concerned with the study of gases, which are indeed fundamental to physics, chemistry and physiology.

The physicist can express their compression by mechanical forces and their expansion by heat in terms of very simple laws, which, when mathematically analysed, lead to the great generalisation of the Kinetic Theory.

The chemist finds in them the simplest kinds of matter, elements and simple compounds which take part in a great variety of changes, and without an understanding of which theoretical chemistry could not exist.

The physiologist could not understand the fundamental processes of life,—respiration and photosynthesis,—until the common gases had been characterised and their relations demonstrated.

The discovery of the common gases and their nature occupied rather less than two centuries, from the time of Galileo to that of Lavoisier.

The Primitive idea of a Gas

From the beginning of the study of matter until the seventeenth century the air was regarded as something not quite material in the sense that earth or water was material. Some ancient Greek philosophers made no clear distinction between air and 'spirit'. They had a conception which has vanished to-day, that of 'breath'. Mist, vapour, breath, the 'spirits' (p. 36) influences (p. 61), and sometimes even the soul, were all thought of as varying degrees of condensation of the same thing, for which we have no word to day, but which they called *pneuma* or *spiritus*. The mediæval method of dealing with *spiritus* was not to try and collect it, for it was thought of as subtle,—capable of penetrating solids,—but to condense it to a liquid form. Thus the active principle, as it were the soul, of wine was evolved by heating it and condensed as *spirits of wine*,—alcohol.

The Physicists and Gases

The first people to study the properties of air were physicists. The Hellenistic scientists, Heron and Philon (p. 33) made a beginning, but it was not until the late sixteenth century, when their works were printed, that these studies began again. J. B. della Porta has experiments on the expansion of gases, much like Heron's, but the first quantitative studies were those of Galileo (pp. 94, 101). From him came the idea of using the expansion of air to measure 'heat', of weighing air and of trying to obtain a vacuum. Torricelli, his disciple, first produced a vacuum: von Guericke (p. 95) made the first air-pump. In all these experiments air is treated as a substance as material as any other, and the 'spiritual' conception is ignored.

An equally important step was taken when Boyle made measurements upon gases and discovered that the volume of air was inversely proportional to the pressure upon it, e.g. that if the pressure were doubled, the volume would be halved, one of the earliest quantitative observations in this field. Boyle was inclined to regard gases as made up of atoms which, like springs or locks of wool, were compressible and elastic, though he saw that the effect could also be explained if they were supposed to be elastic particles in motion.

The Chemists and Gases

We owe the word 'gas' to J. B. van Helmont, about 1620. The extracts given on pp. 168–69, show that he meant by it a sort of subtle spirit readily acted on by the influence of the stars, and the means of causing the phenomena of weather; and he identifies this with a 'wild spirit' produced in chemical reactions—that which bursts the vessels in which it is produced. He did not think it could be collected, but he realised that there were different kinds of gases: thus he distinguished the gas given off when nitric acid acts on metals, from that produced by fermentation and that produced by burning sulphur: indeed he describes fifteen gases,—not all, of course, different chemical individuals. Boyle, in 1660, was the first to collect a gas (hydrogen) by inverting a bottle filled with dilute sulphuric acid in a dish containing the same acid and iron nails, but the idea of collecting gases was not followed up in that century.

Experiments with the air-pump gave Boyle evidence that ordinary combustion would not occur without air, but that mixtures of combustibles with nitre, such as gunpowder, could burn in a vacuum. He thus started the idea that nitre and air had something in common. His experiments on heating metals in closed vessels containing air to discover the reason why they gained in weight led him to no certain conclusions; indeed he supposed that 'particles of fire' passed through the glass to the metal and so increased its weight. Hooke in 1665 also concluded there was a common constituent in air and saltpetre and that this 'dissolved' combustible bodies: he showed very clearly that air was necessary for combustion. John Mayow in 1674 published experiments which showed that in combustion and respiration only a part of the air was used up, and that the same part. He thus proved that air consisted of two parts, one of which was a 'nitro-aerial spirit', while the other was inert. He also showed that red arterial blood gave off bubbles of gas in a vacuum whereous venous blood would not. To us to-day this work seems extremely near to the discovery of oxygen and the true theory of combustion and respiration, and it seems very strange that it should have suffered a century's neglect. This neglect is usually explained by the appearance of the theory of phlogiston, which has been alluded to on pp. 150, 158. This theory appealed to the men of the time, who thought easily in terms of 'spirits', which were as familiar in medicine as in chemistry, but to whom the existence of different kinds of air, as sharply differentiated as e.g. the metals, was wholly unfamiliar. Be this as it may, the phlogiston theory held the field until the seventeen-seventies.

The hypothetical phlogiston was a sort of 'matter of fire', a direct descendant of Aristotle's element of fire. Bodies that could burn were thought to contain much phlogiston: when they burned, it streamed out with a whirling motion. Thus sulphur was taken to be peculiarly rich in phlogiston. When it burned, phlogiston was supposed to stream out of it and the product of its combustion (then supposed to be sulphuric acid) was left. Thus it was supposed that sulphur consisted of sulphuric acid associated with phlogiston. The theory was a hindrance to chemistry in that (as we see in the above case) it made the elements appear to be compounds of their products of combustion with phlogiston, but an assistance in that it gave a general explanation of combustion, which, though wrong, linked together a number of instances of the same type of phenomenon: e.g. combustion and calcination were seen to be the same.

It would seem to us that the fact, known to Boyle and many others, that metals when calcined *increase* in weight, would make the notion that they *lost* phlogiston appear absurd, but some of the supporters of the phlogiston theory supposed that phlogiston (like Aristotle's element of fire) was absolutely light, and that bodies containing it weighed less than they weighed without it.

New Kinds of Gases

The reason why the study of gases proceeded so slowly was the lack of a technique for isolating and handling them. Stephen Hales showed how to collect the 'air' given off when a body was heated (Fig. 40), but he was interested in the magnitude of the volume of 'air' given off when various bodies were heated, not in the kind of air. It was not until after 1750 that different kinds of gases began to be named. Joseph Black, in 1754, was the first to describe a gaseous substance different from air. This was fixed air (carbon dioxide), and he showed that 'magnesia alba' and limestone were compounds of this gas with 'calcined magnesia' and quicklime respectively. In 1766 the eccentric nobleman, Henry Cavendish, published a paper *On Factitious Airs*. He described inflammable air (hydrogen) already collected by Boyle, also fixed air, carbon dioxide. He showed them to be real individuals, not mere varieties of

PLATE XIII

Jacob's Island, Bermondsey. Typical of the housing conditions before 1850. The privies discharge into the river, which is the only source of water for drinking and washing.

PLATE XIV

REPTILE VOLANT d'Eichstadt.

This skeleton had been mistaken for that of an amphibian and for that of a mammal.
Cuvier in 1810 showed that its bone structure was that of a flying reptile, to which he
gave the name Ptérodactyle. (Cuvier. *Recherches sur les Ossemens Fossiles*, 1821–4.)

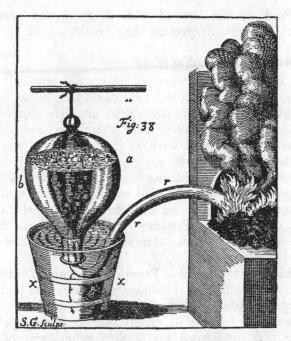

FIG. 40,—The first illustration of an apparatus for collecting a gas. (Stephen Hales. *Vegetable Staticks.* 1737.)

air. He gives practical details of the manner of collecting these gases and transferring them from vessel to vessel. It was then quite clear that there were different kinds of gases. In the next eight years Carl Wilhelm Scheele in Sweden, and Joseph Priestley in England discovered a great number of different kinds of 'air'. Scheele discovered oxygen, chlorine, and silicon fluoride: Priestley discovered oxygen independently, and also ammonia, sulphur dioxide, carbon monoxide, nitrogen, nitrous oxide, nitric oxide, nitrogen dioxide. The gases which we know to be contained in air and water had now been isolated, but their discoverers had not drawn the correct conclusions from their work. Priestley, who discovered oxygen, and Cavendish, who discovered that oxygen and hydrogen combined to form water, were both led by the phlogiston theory and by the age-old idea that air and water were elements, into misinterpreting their results.

Lavoisier and the Composition of Air

After Priestley had discovered the gas we now call oxygen, he was greatly puzzled as to how it could be a better supporter of combustion and respiration than air. He supposed that it was air containing very little phlogiston and that combustibles could therefore burn in it, i.e. transfer their phlogiston to it, with great ease,—so converting it into air fully combined with phlogiston: thus on Priestley's theory (air − phlogiston) was the gas we call oxygen and (air + phlogiston) was the gas we call nitrogen. Scheele, who discovered oxygen before Priestley but announced the discovery later, supposed that it was identical with the 'matter of heat' deprived of phlogiston.

Lavoisier in 1774 heard of Priestley's making of oxygen from *mercurius precipitatus per se* (i.e. mercuric oxide). He had already been experimenting with the heating of metals in air and had found that they *increased* in weight and absorbed a part of the air. He then performed his famous experiment of heating mercury in a limited volume of air (p. 170). There appeared on the mercury in the retort red specks and scales of *mercurius precipitatus per se* and the original 50 cubic inches of air contracted to 42 cubic inches of *moféte*, 'mephitic air' (chiefly nitrogen). When the red matter was collected and heated, 8 cubic inches of 'vital air' were obtained and this mixed with the 42 cubic inches of mephitic air reconstituted 50 cubic inches of common air. Lavoisier concluded then that air was a mixture of vital air (later termed oxygen) and mephitic air (later termed nitrogen); that combustion was chemical combination with vital air and that phlogiston had no real existence. Lavoisier's revival of Boyle's definition of an element as "the ultimate point that analysis can reach" was another great advance. On this basis a list of elements could be made and thenceforward chemical compounds were expressed by him as combinations of known weighable elements. He invented the chemical nomenclature from which our modern one has sprung, and divested chemistry of the ideas and fancies that derived from the Greek doctrine of *pneuma* and the four transformable elements.

Lavoisier and the composition of water

Henry Cavendish proved in 1784 that water was formed when inflammable air (hydrogen) and dephlogisticated air (oxygen) were mixed and exploded by an electric spark. He was somewhat confused by the simultaneous production of a little nitric acid, actually derived from accidental impurities, and also by his firm belief in phlogiston and the elementary nature of water. He therefore supposed water to be pre-formed in each of the gases, so that the combination of the two gases in his view was simply:—

dephlogisticated air   +   inflammable air   =   water
(water − phlogiston)   (water + phlogiston)

Priestley and Watt were also working on this subject. Lavoisier heard of Cavendish's work, hastily repeated it in a much inferior manner and at once published his experiments and the interpretation he gave them. To his discredit he did not acknowledge his source of information, as nine years before he had not acknowledged his debt to Priestley's discovery of oxygen. But he saw the real significance of Cavendish's work and announced that water was composed weight for weight of inflammable air and vital air. This removed the last difficulties in the rejection of the theory of phlogiston and initiated a new era in chemistry.

The significance of oxygen for biology

According to the phlogiston theory, the function of air in the animal economy was to remove phlogiston. Lavoisier showed that the oxygen of the air oxidised the carbonaceous materials of the body producing carbon dioxide, water and heat. He supposed that this took place in the lungs and it was only in 1837 that Magnus proved that oxidation went on all over the body. Priestley in 1778 showed that air vitiated by respiration or combustion could be restored by green plants; in the next year Ingen-Housz showed that light was necessary for this to occur and Senebier in 1782 showed that fixed air (carbon dioxide) was converted into dephlogisticated air (oxygen) in this process.

Thus by 1790 the nature of the two fundamental processes of biochemistry, respiration and photosynthesis, had been explained, and a foundation thus laid for a scientific physiology.

Extracts Concerning Gases

VAN HELMONT ON GAS

He discusses the manner in which the visible vapours from water assume in the air an invisible subtle form, and continues:—

(p. 69) " . . . But because the water which is brought into a vapour by cold, is of another condition than a vapour raised by heat; therefore by the Licence of a Paradox for want of a name, I have called that vapour, Gas, being not far severed from the Chaos* of the Auntients. In the meantime, it is sufficient for me to know, that Gas, is a far more subtle or fine thing than a vapour, mist or distilled Oyliness, although as yet, it be many times thicker than Air. But Gas it self, materially taken, is water as yet masked with the Ferment of composed bodies. . . ."

(p. 74) " . . . It is sufficient that I have known an exhalation arising from beneath, to wit, a vapour, and Gas to be the material cause of every Meteor.† It sufficeth to have known Blas to be the effective cause. . . .

. . . Therefore the live coal, and generally whatsoever bodies do not immediately depart into water, nor yet are fixed, do necessarily belch forth a wild spirit or breath. Suppose thou, that of 62 pounds of Oaken coal, one pound of ashes is composed: Therefore the

* Primitive confused substance from which all things derived.

† Meteor; not a shooting star, but any aerial phenomenon. The usage is preserved in our term Meteorology. *Blas* was a sister-conception to *Gas* and represents something like the influence of the planets.

61 remaining pounds are the wild spirit. . . . I call
this Spirit, unknown hitherto, by the new name of Gas,
which can neither be restrained by Vessels, nor reduced
into a visible body. . . .

(*Oriatrike or Physick Refined.* J. B. van Helmont,*
Tr. by J. C. London 1662.)

BOYLE AND HALES MAKE OXYGEN
WITHOUT KNOWING IT

" . . . and that there is good store of air added to the Minium,†
I found by distilling first 1922 grains of Lead from whence I
obtained only seven cubick inches of air; but from 1922 grains,
which was a cubick inch of Red Lead, there arose in the like
space of time 34 cubic inches of air. . . .

It was therefore doubtless this quantity of air in the Minium
which burst the hermetically sealed glasses of the excellent
Mr. Boyle when he heated the Minium contained in them
by a burning-glass."

(*Vegetable Staticks* by Stephen Hales. London. 1727. p. 287.)

PRIESTLEY DISCOVERS OXYGEN

" . . . At present I am chiefly intent on my experiments, and
I was never more successful than I have been of late. . . .

I have now discovered an air five or six times as good as
common air. I got it first from *mercurius calcinatus per se*, red
lead, &c; and now, from many substances, as quicklime (and
others that contain little phlogiston) and spirit of nitre‡, and
by a train of experiments demonstrate that the basis of our
atmosphere is spirit of nitre. Nothing I ever did has surprised
me more, or is more satisfactory."

(Letter from Joseph Priestley to Rev. W. Turner. April 6, 1775.
Life and Correspondence of Joseph Priestley. J. T. Rutt. 1831. 1.
267.)

* This is a translation of van Helmont's works as edited by his son. It
may not always truly reflect his opinions.
† Red Lead, Pb_3O_4.
‡ Concentrated nitric acid.

SCHEELE PREPARES OXYGEN

I took a glass retort which was capable of holding eight ounces of water, and distilled fuming acid of nitre according to the usual method. In the beginning the acid went over red, then it became colourless, and finally all became red again; as soon as I perceived the latter, I took away the receiver and tied on a bladder, emptied of air, into which I poured some thick milk of lime in order to prevent the corrosion of the bladder. I then proceeded with the distillation. The bladder began to expand gradually. After this I permitted everything to cool, and tied up the bladder. Lastly I removed it from the neck of the retort. I filled a bottle, which contained ten ounces of water, with this gas. I then placed a small lighted candle in it; scarcely had this been done when the candle began to burn with a large flame, whereby it gave out such a bright light that it was sufficient to dazzle the eyes. I mixed one part of this air with three parts of that kind of air in which fire would not burn; I had here an air which was like the ordinary air in every respect. Since this air is necessarily required for the origination of fire, and makes up about the third part of our common air, I shall call it after this, for the sake of short-ness, Fire Air; but the other air which is not in the least service-able for the fiery phenomenon, and makes up about two-thirds of our air, I shall designate after this with the name already known, of Vitiated Air.

(From *Collected Papers of C. W. Scheele*, translated by Leonard Dobbin. G. Bell and Sons. 1931.)

LAVOISIER'S DECOMPOSITION OF AIR

" . . . I took a flask (matras) A of about 36 cubic inches capacity, of which the neck BCDE was very long and had an internal width of about six or seven lines. I bent it as repre-sented in figure 2: in such a way that it could be placed on a furnace MMNN, while the extremity E of its neck would go beneath the bell-jar FG, placed in a bath of mercury RRSS. I introduced into the retort four ounces of very pure mercury, then by sucking with a siphon which I introduced under the bell-jar FG, I raised the mercury as far as LL: I marked this height carefully with a band of gummed paper, and I observed the barometer and thermometer exactly.

Having prepared these things, I lit the fire in the furnace MMNN and kept it going continuously for twelve days in such a way that the mercury was heated almost to the degree required to make it boil. . . . On the second day I began to see swimming on the surface of the mercury some small red particles, which in four or five days increased in volume and number, after which they stopped increasing and remained in exactly the same state. At the end of twelve days, seeing that the calcination of mercury made no more progress, I put out the fire and let the vessels cool. The volume of air contained in the retort and in its neck and under the empty part of the bell-jar, reduced to a pressure of 20 inches and at 10 degrees of the thermometer was before the operation about 50 cubic inches. When the operation was finished the same volume at the same pressure and temperature, was found to be only 42 to 43 inches: there was therefore a diminution of volume of about a sixth. On the other hand, having carefully collected the red particles that had formed and having separated them as far as possible from the running mercury with which they were bathed, their weight was found to be 45 grains. . . .

Fig. 2

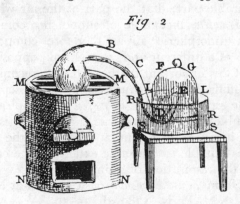

FIG. 41.—Apparatus used by Lavoisier for the decomposition of air.

The air which remained after this operation and which had been reduced to five-sixths of its volume, by the calcination of mercury, was no longer fit for respiration nor for combustion, for animals introduced into it expired in a few moments, and flames were at once extinguished as if they had been plunged into water.

On the other hand, I took the 45 grains of red matter which was formed during the operation, introduced them into a very small glass retort to which was adapted an apparatus suitable for receiving the liquid and aeriform products which might be separated: having lit the fire in the furnace I observed that as the red substance was heated, so its colour increased in intensity. When it finally reached incandescence, the red matter began gradually to decrease in volume, and in a few minutes it had entirely disappeared, at the same time there condensed in the little receiver $41\frac{1}{2}$ grains of running mercury and there passed into the bell 7 or 8 cubic inches of an elastic fluid much more suited than air to support combustion and the respiration of animals. . . .

(He describes tests that showed it to be Priestley's dephlogisticated air, which Lavoisier then called vital air.)

. . . Reflecting on the circumstances of this experiment it is seen that mercury when calcined absorbs the healthful and respirable part of the air, or, to speak more exactly, the base of this respirable part; that the part of the air which remains is a kind of *moféte*, incapable of supporting combustion and respiration: atmospheric air is therefore composed of two elastic fluids of a different and, so to speak, opposite nature.

A proof of this important truth, is that on recombining these two elastic fluids that have thus been obtained separately, that is to say, the 42 cubic inches of *mofète* or irrespirable air, and the eight cubic inches of respirable air, there is once more formed air, in all respects similar to that of the atmosphere, and which is suitable to the same extent for the calcination of metals and the respiration of animals.

(Translated from *Traité Élémentaire de Chimie*. Par M. Lavoisier. Paris. 1789. I. 35–39.)

CHAPTER ELEVEN

The Age of Steam

Productivity

It is obvious that a man aided by machinery can produce a far greater quantity of goods than can the same man equipped only with simple tools. The machine, if hand- or foot-driven, enables him to apply his strength to the best advantage and to make more rapid and precise movements: and if the machine is power-driven, the quantity of goods the worker can produce in a given time is limited only by the amount of 'minding' that the machine requires, that is to say, by the ingenuity and resources of its designer. The extensive use of power-driven machinery created the modern world, in which an enormous variety of machine-made goods are available to all, and distinguishes it from the world before 1800 in which manufactured goods were few and, reckoned in terms of workers' wages, relatively expensive.

The rise of machinery

There were remarkably few machines of any kind before the middle of the eighteenth century. Of those that were in use, some had been known from antiquity, for example, looms, cranes, pumps, lathes, flour-mills, presses, and the potter's wheel; but none of those were more than very simple devices, and, with the exception of flour-mills and, occasionally, pumps, they were generally worked by hand.

The only sources of power before the seventeen-eighties were the muscles of men and animals, water-wheels, and windmills. Thus we noted in chapter IX (p. 159) that the machine, which Rumford in 1798 describes as being used for boring cannon, was turned by horses. Windmills, horse-mills and water-mills did very well for grinding corn, but there was one heavy piece of work for which they were not adequate; namely, the raising of water. In the sixteenth and seventeenth centuries

FIG. 42.—Ball-and-chain pump operated by a tread-mill; from
Agricola's *De Re Metallica*. (1553.)

the population increased, and became more refined and cleanly in its habits, though very much less so than ourselves; consequently large towns began to require a water-supply. Sometimes this could be brought by gravity from higher ground, but sometimes it had to be pumped from a river.

Increasing wealth and more complex ways of living created a demand for more metal: the smelting of metals led to the cutting down of much of our forests and so created a further demand for coal. Thus mining became an important industry in the sixteenth and seventeenth centuries. Mines were driven deeper in search of coal and ore, and in most of them springs were encountered; it was therefore necessary to pump the water out of the workings. A small concern which had to keep a dozen men continuously at the pumps found their wages a heavy charge; and as it was not often possible to use water-mills to work the pumps, the mine-owners were ready to adopt steam-driven pumps as soon as they were available.

The first practical steam-engine for pumping water was invented by Thomas Savery in 1698. It is shown in Pl. XI. Steam generated in a small strong boiler L is blown into an oval copper vessel (p. 181). When the supply of steam is cut off, it condenses; and atmospheric pressure opens the valve at the base of the vessel and forces the water from the mine-sump into that vessel. The steam is once more admitted to the vessel and thus forces the water to the surface. Two vessels were used so that while water was entering one it was being forced out of the other. There is every reason to suppose it was a successful machine, but probably extravagant in fuel and somewhat dangerous, for there was necessarily a pressure of 30–50 lbs. sq. in. in the boiler, which had no safety valve.

The next improvement was the Newcomen engine which held the field for sixty or seventy years. It was safe because it did not operate by the pressure of the steam but by the pressure of the air, and consequently the pressure of the steam in the boiler was kept low. The engine was reliable and was capable of raising great quantities of water, albeit by great expenditure of coal.

The engine (Pl. XII) consisted of a rocking-beam, on one end of which hung the rod and piston of the pump, and on the other the piston of the steam cylinder. These were made to balance each other by counterpoises so that a very slight pressure of steam would push up the piston of the steam cylinder. Steam was generated in a low-pressure boiler and was admitted by a valve (automatic or hand-operated) to the cylinder, which it filled, so raising the piston. The valve was now closed and a jet of cold water admitted, through the valve and pipe below the cylinder. The cold water condensed the steam and the atmosphere pressed down the piston with a force that might amount to several tons. This pulled down the rocking-beam and raised the pump-rod and piston, so forcing the water to the surface. This type of engine was in general use for many years, but was very wasteful because at every stroke steam had to be used to heat the whole cylinder and piston from the temperature of the condensing water up to the temperature of the steam. Wastefulness did not matter very much in coal-mines where fuel was cheap, but in the Cornish tin-mines, where all the fuel had to come by sea and be carted from ports, it was a serious matter.

The earlier steam-engines, it should be noted, were pumps, and were not capable of rotating a wheel or shaft, but in Plate XII is shown a Newcomen engine which is being used to turn a wheel, probably fairly late in the eighteenth century. The slow-moving Newcomen engine was not, however, at all suitable for such work.

Power-driven Machinery

We have seen that in the sixteenth and seventeenth centuries there were few machines of any kind, but the eighteenth century saw the invention of machinery for spinning of yarn. England had for centuries been a cloth-manufacturing country: in the eighteenth century the population began to increase rapidly, and at the same time new markets were opened in all the lands to which our sea-borne trade penetrated. Traders and merchants began to find that they could sell more cloth than the country could produce. The slowest process in the making of cloth was spinning; the weavers, dyers, etc., could have handled more yarn than was being made, so there was a strong need for machinery to speed up spinning, and between 1770 and 1780 were invented a number of spinning-machines, capable of being driven by power. At first the only suitable power was that of the water-wheel, and so the spinning-machines were congregated in "mills", usually in some Yorkshire or Lancashire valley where there was a swiftly-flowing stream. This was the beginning of the factory-system, which brought about the greatest of revolutions in human ways of living. One source of power drives many machines, so these must be congregated in one building, and the worker instead of weaving or spinning at home must spend his day in the factory. Before 1770 the great majority of other than agricultural workers worked at home, now almost everybody works in a factory—a change, we may suppose, that is greatly for the worse.

Water-wheels are stopped by severe frost, or by drought or flood, and there were not many places in Great Britain where water-power was available. So textile machinery created a demand for an engine that could act as prime-mover for a factory.

The Engines of James Watt

The fuel consumption of the Newcomen engine was very high, and Watt was the first to grasp the necessity of economising heat; he was also the first to make steam-engines which were suitable for turning a wheel instead of pulling a pump-rod. Watt's principles, which changed the steam-engine from a mining appliance to a world-power, were embodied in patents of 1769, 1781 and 1782. The most essential were as follows:

1. *The cylinder must be kept as hot as the steam that enters it.*

The Newcomen engine condensed the steam in the cylinder, which therefore at each stroke had to be heated by the steam from, say, 20° C. to 100° C. This heating of the cylinder walls condensed much steam without getting any useful work from it, and indeed this type of engine consumed at each stroke about four times as much steam as was needed to fill the cylinder. Watt avoided this necessity by surrounding the

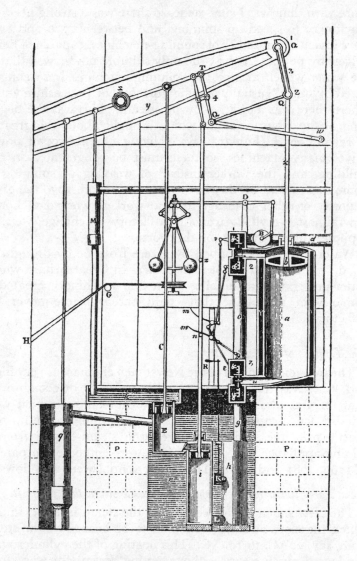

FIG. 43.—Part of early Watt engine. Note (*a*) lagged cylinder for double action; (QQ) parallel-motion by which piston-rod can remain vertical; (*zz*) governor; (*h*) separate condenser; (E, *i*) pumps to maintain vacuum and remove water from condensers: pump-rod operates steam-valves.

cylinder (Fig. 43, *a*) with non-conducting material and by applying his second principle, namely, that,

2. *Condensation is to take place in a separate vessel, the condenser* (*h*), *connected to the cylinder through a cock or valve and kept free from accidentally-entering air by a pump.*

The cylinder was thus kept continually hot and the condenser continually cold; so heat was not wasted in repeatedly changing the temperature of parts of the engine. His third principle was of even greater importance. Instead of using steam merely to make a vacuum in the cylinder, he also used its pressure to impel the piston. He therefore suggests:

3. *Employment of the expensive force of steam and, if necessary, dispensing with condensation.*

The engine with these improvements was much more efficient. In 1781 Watt used devices to convert the back-and-forth motion of the piston into the rotary motion of a wheel. He did not at first employ the familiar method of the crank, for a rival had patented this.

In 1782 and shortly after, he introduced two most important advances. First was the use of double action, i.e. the application of steam on one side of the piston while a vacuum was applied to the other; this doubled the duty done by each cylinder. Secondly he introduced the plan of stopping the admission of steam when the piston had travelled over but a part of its course and then allowing the expansion of the compressed steam to complete the stroke. This effected considerable economy. Finally he provided for steady running by the use of the throttle valve (B) and governor (*z z*) which automatically cut down the steam supply to the cylinder if the rate of running of the engine increased—e.g. on account of the lightening of its load—and *vice versa*.

Watt may be said to have created the steam-engine as a source of power. His purpose had been to make a more economical pumping-engine, but a far more important result was the use of the new engines, first to drive machinery, secondly to accelerate sea- and land-transport.

The Factories

As we have seen, spinning- and carding-machinery was already invented and waiting for a prime mover, and so it was not long before Lancashire and the West Riding were covered with factories and 'mills' in which a large number of textile machines were driven by a single steam-engine. Soon after the turn of the century the power-loom was introduced, and by then the greater part of the textile trade was mechanised after a fashion.

The effects were far-reaching. First of all the workers who had spun and woven in their cottages were congregated in large factories, and the factories were grouped in the places where coal and water and labour were most easily come by. The result was the migration of the working-people from villages, and the formation of towns of a size hitherto unknown. There was no attempt to regulate building or to provide adequate water-supply or sanitation, so that the workers were horribly over-crowded in damp ill-ventilated houses, under conditions of terrible filth (pp. 201–204). The factory, moreover, did not require the skilled labour of the former spinners and weavers; in their place it needed a few engineers and fitters and a vast number of the cheapest unskilled workers, whose task was un-endingly to push wool into a machine or guide a piece of cloth through it, or tie threads when they broke, and the like. There were no factory- or labour-regulations, and the employers hired the cheapest form of labour—children and preferably orphan children apprenticed by workhouses. They were incredibly overworked, underfed and cruelly beaten: their death rate is unknown but must have been gigantic, for even in the eighteen-forties, when conditions had somewhat improved, six out of every ten people who died in the textile districts were under twenty—the proportion today is one in eight. Legislation to prevent this wholesale slaughter of the innocent proved exceedingly difficult to pass, for the workers were quite un-represented in Parliament; and it was not until 1847 that any really effective measures were taken to restrict the employment of young children. The new steam-engines and the railways created a further great demand for coal which further increased the congestion of the Black Country. Women and children were employed in the mines in even more horrible conditions than in the factories (pp. 187–89).

The goods were turned out. England sold textiles and engines and rails to the world: fortunes were made and re-invested, and by the eighteen-fifties she had founded a great age of industrial prosperity on the blood and misery of the exploited workers. The years from 1780 to 1850, the most brilliant in engineering progress, are the most shameful of our history from the point of view of social morality.

Mechanisation outside the textile trade

During the first half of the nineteenth century, there was surprisingly little use of machinery in other trades than the textile. Power was used almost exclusively for work which was too heavy for men to perform economically. Thus paper-mills, saw-mills, rolling-mills, pumping-plant were commonly driven by steam-engines, but the speeding up of output by the use of light workshop machinery did not make much headway until the gas-engine and later the electric motor provided useful small units of intermittent power.

Examples Concerning the Early Use of Power

SAVERY'S ENGINE

The manner of working the engine

The first thing is to fix the engine in a good double furnace, so contrived that the flame of your fire may circulate round and encompass your two boilers to the best advantage, as you do coppers for brewing. Before you make any fire, unscrew g and n, being the two small gauge-pipes and cocks belonging to the two boilers, and at the holes, fill l, the great boiler, two-thirds full of water, and d, the small boiler, quite full; then screw in the said pipes again as fast and tight as possible; then light the fire at b. When the water in l boils, the handle of the regulator, marked z, must be thrust from you as far as it will go, which makes all the steam rising from the water in *l* pass with irresistible force through *o* into *p*, pushing out all the air before it, through the clack *r*, making a noise as it goes; and when all is gone out, the bottom of vessel *p*, will be very hot; then pull the handle of the regulator to you, by which means you stop *o*, and force your steam through *o* into *p*, until that vessel has discharged its air through the clack *r*, up the force-pipe. In the meantime, by the steam's condensing in the vessel *p*, a vacuum or emptiness is created, so that the water must, and will, necessarily, rise up, through *t*, the sucking-pipe, lifting up the clack *r*, and filling the vessel *p*.

In the meantime, the vessel *p*, being emptied of its air, turn the handle of the regulator from you again, and the force is upon the surface of the water in *p*, which surface being only heated by the steam, it does not condense it, but the steam gravitates or presses with an elastic quality like air; still increasing its elasticity or spring, till it counterpoises, or rather exceeds the weight of the water ascending in *s*, the forcing-pipe, out of which, the water in *p* will be immediately discharged when once gotten to the top, which takes up some time to recover that power; which having once got, and being in work, it is easy for anyone that never saw the engine, after half an hour's experience, to keep a constant stream running out the full bore of the pipe *s*; for, on the outside of the vessel *p*, you may see how the water goes out, as well as if the vessel were transparent; for, as far as the steam continues in the vessel, so far is the vessel dry without, and so very hot, as scarce to endure the least touch of the hand; but as far as the water is, the said vessel will be cold and wet, where any water has fallen on it; which cold and moisture vanishes as fast as the steam, in its descent, takes place of the water; but if you force all the water out, the steam, or a small part thereof, going through *r*, will rattle the clack, so as to give sufficient notice to pull the handle of the regulator to you, which, at the same time, begins to force out the water from *p*, without the least alteration of the steam; only, sometimes, the stream of the water will be somewhat stronger than before, if you pull the handle of the regulator before any considerable quantity of steam be gone up the clack r; but it is much better to let none of the steam go off, for that is but losing so much strength, and is easily prevented, by pulling the regulator some little time before the vessel forcing is quite emptied. This being done, immediately turn the cock or pipe of the cistern x, on p, so that the water proceeding from x, through y, which is never open but when turned on p, or P; but when between them is tight and stanch; I say, the water, falling on p, causes, by its coolness, the steam, which had such great force just before to its elastic power, to condense and become a vacuum or empty space, so that the vessel p is, by the external pressure of the atmosphere, or what is vulgarly called suction, immediately refilled, while p is emptying; which being done, you push the handle of the regulator from you, and throw the force on p, pulling the condensing pipe over p, causing the steam in that vessel to condense, so that it fills, while the other empties. The labour of turning these two parts of the engine, viz., the regulator and water-cock, and attending the fire, being no more than what a boy's strength can perform for a day together, and is as easily learned as their driving of a horse in a tub-gin; yet, after all, I would have men, and those, too, the most apprehensive, employed in working of the engine, supposing them more careful than boys. The difference of this charge is not to be mentioned or accounted of when we consider the vast profit which those who use the engine will reap by it.

(*The Miner's Friend*. Thomas Savery. 1702.)

CHILD LABOUR IN THE WOOLLEN MILLS: 1832

What intervals have these children allowed for their meals; for breakfast, for instance?—I have been in mills at all hours, and I never in my life saw the machinery stopped at breakfast time at any of the mills.

How do they get their breakfast?—They get their breakfast as they can; they eat and work, there is generally a pot of water porridge, with a little treacle in it, placed at the end of the machine, and when they have exerted themselves to get a little forward with their work, they take a few spoonfuls for a minute or two, and then to work again, and continue to do so until they have finished their breakfast. This is the general practice, not only of the children, but of the men in the woollen mills in this district.

Is there any allowance of time for the afternoon refreshment, called drinking?—No allowance for drinking, more than breakfast.

How much at dinner time, as far as you have been able to judge, speaking now of the summer season?—In summer some of the mills allow an hour for dinner.

And some less?—Some less; some 40 minutes.

What is the usual time allowed in winter?—There is no time allowed in winter, only just sufficient to eat their dinner; perhaps ten minutes or a quarter of an hour; and in some cases they manage the same at noon as they do at breakfast and drinking.

The children are employed as what they call pieceners, are they not?—Yes.

Does not that require them to be constantly on their feet?—Always on their feet when at work; they cannot sit and piece.

During the hours of which you speak, then, the children are to be constantly on their feet, excepting these very short intervals?—Yes.

Only this very short time for dinner?—That is the only interval they have for rest, except it may sometimes happen that they may be out of what we call jummed wool, and then the children have a short opportunity to rest themselves, and even then they are frequently employed in cleaning the cording machine.

You say you have observed these children constantly for many years going there early in the morning to their work, and continuing at it till so late at night?—Yes; I have seen children during the last winter coming from work on cold dark nights between 10 and 11 o'clock, although trade has been so bad with some of the mills that they have had nothing to do; others have been working seventeen or seventeen and a half hours per day.

This requires that the cottagers should wake their children

very early in the morning?—It cannot be expected they can go to their work asleep.

How early do you think that they leave their homes?—I can tell you what a neighbour told me six weeks ago; she is the wife of Jonas Barrowcliffe, near Scholes; her child works at a mill nearly two miles from home, and I have seen that child coming from its work this winter between 10 and 11 in the evening; and the mother told me that one morning this winter the child had been up by two o'clock in the morning, when it had only arrived from work at eleven; it then had to go nearly two miles to the mill, where it had to stay at the door till the overlooker came to open it.

This family had no clock, I suppose?—They had no clock; and she believed, from what she afterwards learnt from the neighbours, that it was only two o'clock when the child was called up and went to work; but this has only generally happened when it has been moonlight, thinking the morning was approaching.

Is this practice general in the entire neighbourhood?—It is the general practice of the neighbourhood; and any fact that I state here can be borne out by particular evidences, that, if required, I can point out.

What has been the treatment which you have observed that these children have received at the mills, to keep them attentive for so many hours at such an early age?—They are generally cruelly treated; so cruelly treated, that they dare not hardly for their lives be too late at their work in the morning. When I have been at the mills in the winter season, when the children are at work in the evening, the very first thing they inquire is, "What o'clock is it?" if I should answer "Seven", they say, "Only seven! it is a great while to ten, but we must not give up till ten o'clock or past." They look so anxious to know what o'clock it is, that I am convinced the children are fatigued, and think that even at seven they have worked too long. My heart has been ready to bleed for them when I have seen them so fatigued, for they appear in such a state of apathy and insensibility as really not to know whether they are doing their work or not; they usually throw a bunch of 10 or 12 cordings across the hand, and take one off at a time; but I have seen the bunch entirely finished, and they have attempted to take off another when they have not had a cording at all; they have been so fatigued as not to know whether they were at work or not.

Do they frequently fall into errors and mistakes in piecing when thus fatigued?—Yes; the errors they make when thus fatigued are, that instead of placing the cording in this way (*describing it*), they are apt to place them obliquely, and that causes a flying, which makes bad yarn; and when the billy-spinner sees that, he takes his strap or the billy-roller, and says, "Damn thee, close it—little devil, close it," and they smite the child with the strap or the billy-roller.

You have noticed this in the after part of the day more particularly?—It is a very difficult thing to go into a mill in the latter part of the day, particularly in winter, and not to hear some of the children crying for being beaten for this very fault.

How are they beaten?—That depends on the humanity of the hubber or billy-spinner; some have been beaten so violently that they have lost their lives in consequence of their being so beaten; and even a young girl has had the end of a billy-roller jammed through her cheek.

What is the billy-roller?—A heavy rod of from two to three yards long, and of two inches in diameter, and with an iron pivot at each end; it runs on the top of the cording over the feeding cloth. I have seen them take the billy-roller and rap them on the head, making their heads crack, so that you might have heard the blow at the distance of six or eight yards, in spite of the din and rolling of the machinery; many have been knocked down by the instrument. I knew a boy very well, of the name of Senior, with whom I went to school; he was struck with a billy-roller on the elbow, it occasioned a swelling, he was not able to work more than three or four weeks after the blow, and he died in consequence. There was a woman in Holmfirth who was beaten very much, I am not quite certain whether on the head, and she also lost her life in consequence of being beaten with a billy-roller. This which is produced (*showing one*) is not the largest size; there are some a foot longer than that;

it is the most common instrument with which these poor little pieceners are beaten, more commonly than with either a stick or a strap.

How is it detached from the machinery?—Supposing this to be the billy-frame (*describing it*), at each end there is a socket open, the cording runs underneath here, just in this way, and when the billy-spinner is angry, and sees the little piecener has done wrong, he takes off this and says, "Damn thee, close it."

You have seen the poor children in this situation?—I have seen them frequently struck with the billy-roller; I never saw one so struck as to occasion its death, but I once saw a piecener struck on the face by a billy-spinner with his hand, until its nose bled very much; and when I said, "Oh dear, I would not suffer a child of mine to be treated thus," the man has said, "How the devil do you know but what he deserved it? What have you to do with it?"

What moral effect do you think it has on the minds of the children who labour thus at this early period of life?—With regard to the morals of the children who work in mills, we cannot expect that they can be so strict as children who are generally under the care of their parents. I have seen a little boy, only this winter, who works at a mill, and who lives within 200 or 300 yards of my own door; he is not six years old, and I have seen him, when he had a few coppers in his pocket, go to a beer shop, call for a glass of ale, and drink as boldly as any full-grown man, cursing and swearing, and saying he should be a man as soon as some of them.

(The above is taken from sworn evidence from the Report from the Committee on the Bill to regulate the Labour of Children in the Mills and Factories of the United Kingdom. 1832. It should be taken as representing the worst conditions of child-labour rather than those commonly prevailing at that time.)

CHILD LABOUR IN THE MINES

Where do you work?—In Walkden Moor, at a place called New Engine.

What is your work?—Drawing from a man.*

Describe it.—We have belts and chains round our bodies, and tubs to them.

So the tub is fastened to you?—We have hooks at the ends of the chains, and so that is fastened to the tub, and we pull it about, bending double in this manner (*bending his head down to the table*).

At what time do you go down into the mine?—Mostly we go down at half-past five, or between that and six.

When do you come up?—We are uncertain; sometimes at twelve, sometimes before, sometimes after six, sometimes we have been down fourteen hours.

When do you eat?—We eat mostly about ten or twelve; we have no time.

Do you stop working when you eat?—Yes; we stop about a minute or so, and sometimes we work and eat together; when the baskets are full we have more time.

Do you ever stop longer?—Sometimes we stop three or four hours, when our baskets are full.

Have you ever worked a whole day together without stopping to eat?—Yes, we have; and sometimes we can't eat at all for want of something to drink.

Don't you ever take bottles of drink down?—No; they are sure to break.

Do you ever take cans?—No; there is no conveniences at all to take anything about; and if they took cans full, they would be drunk by others.

What do you eat for dinner?—Sometimes bread and cheese, and sometimes bread and butter, and sometimes pies of potatoes and flesh; and in fruit time we get fruit pies.

How often do you have flesh pies?—Sometimes we have a month or two without them; I cannot do with them at all, as they come up again from the mouth.

What makes them come up again?—It is bending in this manner.

* Taking the coal cut by a miner to the shaft from which it was raised to the surface. The tub was a heavy box, wheeled or otherwise, which the children dragged by means of the belt and chains.

You mean that you are sick?—Yes.

Are the boys often sick?—Yes, very often.

Do you breakfast in the mine?—No; my breakfast is coffee and butter-cakes at home, and at night we get potatoes and bacon.

How much do you get a week?—The highest I ever got in my life was 6s. a week, but now about 19s. the last month.

Is that the general wages now?—Yes.

Do girls get the same?—About the same.

Do you work generally bent?—Yes, regularly bent; we never stand up.

Are you often ill?—No; I never was plagued with bad health, not so very much.

How old were you when you went into the mine?—Going nine year since.*

At what age do they usually go into the mine?—There is some that is under six years of age.

Do these little children do the same work as you?—What the little ones can't manage himself the others help; there is two to a basket when one cannot manage it. The little ones bend down just as we do; they hooks the chains to the staples in the baskets, and draw them along; when they thrutch they push with their breasts.

Had you pains in your breast with thrutching?—Yes, sometimes.

Do the girls thrutch?—Yes, when they are little, and when they grow bigger they draw by themselves.

Do the boys and girls work exactly in the same way?—Yes, both exactly in the same way.

Do they ever get crooked with bending?—Yes; there is some as grows crooked as goes in pits, and some as does not.

How are they crooked?—Some in their backs and some in their legs; but very few boys crooked any way; when they grow into men, about thirty or forty years of age, then they starten a growing crooked.

Have the children in pits as good health as those out of pits? —Yes; some is fully better.

* He was consequently eight when he began to work.

Are the people who work out of pits ever crooked?—No; it is not the same as when they are regularly bent in this manner.

What is the weight of the basket you draw?—About four hundred weight.

Have sometimes two little children under six such a basket to draw between them?—Yes; in an easy place, where it is near where he teams the coal, one will do it; some of them have 200 yards to draw the baskets, and when their legs martches they stop and stand in this manner with their hands on the basket as well as they can.

Is the greater part of the mine dry?—I don't know any place that is dry, except where I work.

How high is it where you work?—It is about three quarters of a yard, and some is about thirty inches or so.

Do you work barefoot?—Some with clogs and some barefoot, boys and girls the same; they take all their clothes off except breeches; girls wear breeches.

How do the girls go down?—Just the same as the boys, by ladders or baskets.

(First Report of the Central Board of His Majesty's Commissioners . . . as to the Employment of Children in Factories. 28 June 1833.)

CHAPTER TWELVE

Science and Public Health

The increase of longevity

The amazing increase in the health of civilised humanity is best reflected in the figures for the expectation of life of a human being: that is, the probable number of years a new-born baby may be expected to live. If the expectation of life is 50, it is roughly true that as many people live to be more than fifty as die before they are fifty: it does not of course mean that everybody or even most people die at fifty. The following table speaks for itself:

Date	Expectation of life	Figures taken from
Before 1600	8	
1600–1650	13	
1650–1700	27	Records of
1700–1750	31	City of Geneva
1750–1800	40	
1840	45	
1841	41	
1881	45	
1921	57	England: Registrar
1940	63	General's Reports
India, 1930	27	

There appears to have been slow but considerable progress up to about 1800, then very little improvement till 1880, then a very rapid increase.

Why people die

Everyone has to die, but those who die young usually do so from what we may call a germ-disease—that is, the multiplication in the human body of a microscopic organism that may be a virus-particle, a bacterium or a protozoon. A person cannot be infected unless the organism enters his body: he is likely to suffer the more severely, the worse is his state of general health. Bacteria are transferred from the sick to the healthy by:

1. *Sputum droplets* ejected by talking, coughing, sneezing. Very many diseases can be so transmitted. This mode of infection is obviously increased by (*a*) overcrowding, (*b*) lack of ventilation.

2. *Infected water*. The excreta of typhoid and cholera patients swarm with the bacteria that cause these diseases. If these are drained into rivers, the water will infect all who drink it and are susceptible.

3. *Infected food*. Milk is the chief danger. It commonly contains tubercle bacilli and is a ready breeding ground for any bacteria carried by those who handle it. Flies may carry bacteria from excreta to food.

4. *By insects*. Infected insects bite men and inject germs. Plague is carried by the rat-flea, typhus by the louse, malaria and yellow fever by mosquitoes.

5. *Venereal diseases*. Conveyed by sexual intercourse with an infected person.

6. *By infection of wounds*. All natural wounds become infected with bacteria, but certain strains are particularly dangerous. The hands and instruments of surgeons often carried these before antiseptic precautions were discovered.

All these modes of infection are partly or completely preventable, as also are *deficiency* diseases and conditions, arising from insufficient or improper food: e.g. scurvy, rickets, and starvation. These caused few deaths, but much ill-health.

Thus disease is rampant in communities which are ill-housed, whose drinking water contains sewage, who drink dirty milk, are verminous, unchaste, and underfed. All this was true of the majority of town-dwellers in the eighteenth century and earlier. There seems to have been a slow improvement in most of these matters from the sixteenth century onwards. People slowly became more refined in their habits, more averse to gross dirt and vermin, and the standard of housing steadily improved.

This is reflected in the first steady rise of expectation of life, but in the years round 1800 this general improvement was countered by the factory-system which caused an enormous increase in the size of towns and a consequent overcrowding, which was very little diminished till the end of the nineteenth century. The town population was also exposed to infected water in a way that the countryman was not; town-dwellers were moreover generally ill-nourished and overworked. There was a huge unnecessary death-rate from 'fever', which meant typhoid, and also from tuberculosis, which was about eight times as prevalent as it is to-day.

From the eighteen-forties certain enlightened people such as Edwin Chadwick saw clearly that the poor suffered from 'fever' far more severely than the rich, and managed to establish some connection between the presence of fever and the absence of drains. But what was the connection? Here the world remained at a loss until the years round 1880.

Theories of infection

The standard theory of infection, before the germ-theory of disease was accepted, was that of *miasma*. Marshes, decaying matter, sick persons, were thought to give off a miasma, a subtle gaseous poison or influence which caused disease in those that breathed it. The theory indicated that where there was a stink, disease might be expected; relics of this are found in people who expect to be infected by a smelly drain. In fact, stink is harmless in itself: it is merely an indication of bacterial action elsewhere.

The measures employed to remove bad smells, namely, fresh air, and soap-and-water, are excellent removers of bacteria, so some, such as Florence Nightingale, who warred on disease by cleanliness, were in fact rewarded by a decrease of wound-infections and of the droplet-carried and insect-borne diseases; but soap-and-water could have no effect on the water-borne typhoid and cholera.

The germ-theory of disease was found out in the same round-about manner as most great scientific discoveries.

The Work of Pasteur and Lister

Louis Pasteur in 1857 began to study alcoholic fermentation and thence came to investigate the similar phenomena of the souring of milk and the putrefaction or 'going bad' of broths, etc. He observed, as many had done before him, that the process was accompanied by a vast multiplication of microscopic organisms, yeasts, bacteria, etc.; but he differed from his predecessors in concluding that these organisms were the causes of the changes and not a mere incident, and by a series of brilliantly conceived experiments, such as that described on pp. 206–207, he proved, in face of much opposition and controversy, that if micro-organisms were excluded from anything susceptible of fermentation or putrefaction it would not ferment or putrefy. This work was not only of enormous theoretical interest, but of great importance for the brewing-trade of France. In 1862 Pasteur was asked to investigate a disease of silkworms which was ruining the industry, and he proved it to be due to living 'corpuscles' which multiplied in the silkworms. Pasteur then had to abandon this type of work on account of ill-health.

Meanwhile Joseph Lister had been investigating the fearful mortality that attended surgical operations. Anæsthetics had been in use since 1847, which made for slower and more careful surgery, but the mortality from septicæmia was still enormous, and about half his amputation cases died from this cause. Why did wounds become septic, he wondered; and it seemed to him to result from a sort of putrefaction of the fluids in and about the wound. In 1865 a colleague drew his attention to the recent work of Pasteur; putrefaction was due to micro-organisms— the same might be true of sepsis. Lister then designed a technique for excluding micro-organisms from surgical wounds and killing them if present by means of disinfectants, chiefly carbolic acid. He had immediate and brilliant success, and as years went on he gradually discarded the use of these somewhat poisonous and irritating disinfectants, and developed the modern surgical technique of sterilisation of the surgeon's hands

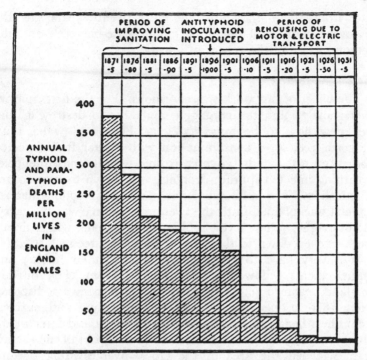

PERIOD OF IMPROVING SANITATION			ANTITYPHOID INOCULATION INTRODUCED ↓			PERIOD OF REHOUSING DUE TO MOTOR & ELECTRIC TRANSPORT						
1871 -5	1876 -80	1881 -5	1886 -90	1891 -5	1896 -1900	1901 -5	1906 -10	1911 -5	1916 -20	1921 -5	1926 -30	1931 -5

ANNUAL TYPHOID AND PARATYPHOID DEATHS PER MILLION LIVES IN ENGLAND AND WALES

FIG. 44.—Typhoid deaths, 1871–1931. (From Sherwood Taylor's *The Conquest of Bacteria*, by courtesy of Messrs. Secker and Warburg.)

and the patient's skin, of instruments and dressings. This work not only decreased the mass of deaths due to wounds, surgical and otherwise, but made it possible to save a great number of lives by surgical operations which would formerly have meant certain death from septic poisoning.

It is very remarkable that neither Pasteur's work on silk-worms nor Lister's on surgical sepsis, nor Davaine's proof that anthrax in animals was due to bacteria, led to the general inference that infectious diseases were caused by bacteria or other micro-organisms. This great step was mainly due to Robert Koch, who worked out the life-history of bacteria and discovered the technique of cultivating them on jelly and making them readily visible under the microscope by staining them with dyes. The germ-theory of disease may be said to date from about 1880 and the next twenty years saw the discovery of most of the important disease-bacteria, and the demonstration of the routes of infection.

Preventive Medicine

The effect of this work was enormous. For the first time we knew how to find the contagion and how to destroy it, and we knew how to be *rationally clean*. Food and water must be germ-free; gross contamination with matter from infected bodies must be avoided; overcrowding, with its consequence of breathing infected sputum-droplets, was shown to be a source of infection. The only important means of infection that remained unexplained until the twentieth century was the carrying of plague, yellow fever, typhus, malaria, etc., which finally were tracked down to the bites of infected insects.

It is very hard to make people change their habits, and improvements followed rather slowly. From 1865 onward England began to get clean water and proper sewage disposal; from 1880 proper isolation of the infectious and surgical cleanliness began to be practised in hospitals and later in the home; from 1900 when the cheap road transport of the electric tram widened the area of the cities, overcrowding became greatly reduced.

These were the great steps towards prevention of disease in England: in tropical countries an equally great step was the control of parasites, more especially the malaria-carrying mosquito.

The Cure of Disease

Until about ten years ago it may be said that, with few exceptions, infectious diseases were left to cure themselves. The doctor prescribed conditions of rest, warmth, diet, etc.: but the body had to fight and destroy the germs by itself. As soon as the germ-theory of disease was evolved, it became clear that if the germs in the body could be killed, the disease would be cured. But with two exceptions, the drugs that were capable of killing bacteria in the test-tube were poisonous to the patient and had little or no effect on the bacteria in his tissues. Yet there was a shining exception. Quinine did kill the malaria parasite without injuring the patient: but the search for drugs effective against other germs remained fruitless for decades.

Inoculation and Vaccination

It has always been known that certain diseases are rarely, if ever, contracted twice in a lifetime. Perhaps the most universal and dreaded of diseases before the nineteenth century was small-pox. The Turks appear to have invented the practice of *inoculation*, that is, of infecting children with matter from a very mild case of small-pox, so causing a very mild attack, which none the less was a lasting protection. The practice was brought to England in the early eighteenth century. At the end of the century (1798) Edward Jenner announced his discovery that an attack of cow-pox protected against small-pox, and so introduced the practice of vaccination. There is to-day almost no mortality from small-pox in countries where vaccination is largely practised. Inoculation with mild strains of living organisms is still practised. Pasteur's treatment of hydrophobia with an 'attenuated' virus, and the B.C.G. treatment designed to protect against tuberculosis is on the same basis. These methods are dangerous, for it is difficult to be certain that the bacteria are of a mild type; and so inoculation with dead bacteria is preferred. If dead bacteria are injected into the body it is stimulated to produce complex substances ('anti-bodies') which are capable of killing bacteria of the same kind and neutralising their poisons, and these anti-bodies protect against any living bacteria that may subsequently enter it. This process has proved effective as a protection against typhoid, cholera, plague and whooping-cough.

A third resource is anti-toxin treatment, in which an animal makes the protective anti-bodies. The dead or living bacteria are injected into, e.g., a horse, which develops an immunity to them: the blood-serum of this horse contains the protective substances and if injected into a human body will destroy the bacterial toxins. This anti-toxin treatment has had great success in combating diphtheria, but although it is of some assistance in many diseases, diphtheria is the only disease for which such treatment is a spectacular success.

Chemotherapy

Chemotherapy is the treatment of a germ-disease by a drug that adversely affects the bacterium or other parasite without having an ill effect on the human body. The effect of quinine on malaria and the effect of mercury on syphilis had been known since the seventeenth century, but no other infectious diseases seemed to be influenced by drugs until the twentieth century, when further progress was made by Ehrlich's discovery in 1910 of an organic arsenical compound, salvarsan, that had a drastic effect on the germ of syphilis. Some tropical diseases were successfully treated with synthetic drugs, but the great medical event of the century has been the discovery of a group of synthetic drugs, known collectively as the *sulphonamide* drugs, which have the most spectacular effect on a number of bacterial diseases. Septic wounds, the infections which too often result in maternal mortality, pneumonia, meningitis, gonorrhœa, are cured with such ease as enormously to reduce the sickness-rate and the mortality from these causes. The latest of these anti-bacterial drugs, penicillin, is not of the sulphonamide group and is obtained from moulds; this material appears to open up further possibilities, and we know of no reason why, as research progresses, every type of bacterial disease should not be remedied by some drug.

Public Health Services

It is not sufficient to *know* the means of preventing and curing disease: it is equally necessary to *provide* these means and even to compel their use. A great part of the public health service consists of seeing that people, who cannot or will not provide or ensure such things for themselves, have sufficient space, light, food, pure water and the means of removing waste matter; and none of these things can be provided by the medical man. Legislation is needed to prevent builders from erecting or people from inhabiting dark, overcrowded, ill-ventilated houses: municipal or state action is needed to provide and control water-supply, and to lay down and connect sewers; last, but not least, a nation with an economic system which allows the unemployed to starve or exist at a low level of nourishment can only be healthy in times of good trade.

The English tradition was that "a man may do what he will with his own"; and accordingly throughout the nineteenth century Parliament was very loth to impose any rules or restrictions on landlords, factory-owners, employers, and vestries,* who effectually controlled the lives of the workers. The beginning of public health services was the adoptive legislation of 1848 giving local authorities powers, which they might use if they chose, to appoint medical officers and reform sanitation. In 1866 they were compelled to take action; in 1871 a very great step was taken, the setting up of a central authority with powers of compulsion—the Local Government Board. In 1919 this was replaced by the Ministry of Health.

Before 1907 there was little provision for free medical services except in the hospitals, and by the traditional charity of doctors. In that year the School Medical Services began and have since greatly increased. The greatest step was, however, the National Health Insurance Scheme, which began in 1911; the Maternity and Infant Welfare Services are still more recent. The system was far from complete for it could not ensure that every citizen did promptly and automatically have medical or surgical treatment without loss or privation; but this seems to be assured by the legislation which came into force in 1948.

The Residual Problem

Our progress towards the ensuring of the maximum possible degree of health for every citizen has gone far but is not complete. It requires better organised medical and social services, but far more does it require better organised medical research. Medical and social services may diminish the virulence and spread of disease, but medical research cuts it off at the root. If the whole wealth of the country had been expended on preventing and treating pneumonia by the methods of 1935, it could not have lowered its death-rate by a quarter of the

* Roughly equivalent to our local authorities.

lowering that resulted from the discovery of sulphapyridine; yet the cost of the research into this was the labour of a handful of men and the expenditure of only a few thousand pounds. In spite of such facts as these only a modest provision is yet (1948) made for medical research.

Lastly, we are to remember that England is not the world. The health of India is worse than that of England in the evil days of the eighteen-thirties—and the Indians are men, like ourselves, to whom we owe the same duty and compassion as we owe to our own nation. The fact that in India, with nearly 400,000,000 people, the expectation of life is only 27, makes the raising of our English expectation of life to 63 at once a small achievement and a reproach to our indifference.

Examples Illustrating the Progress of Public Health

BETTER-CLASS HOUSES IN THE SIXTEENTH CENTURY

As to the floors, they are usually made of clay covered with rushes, that grew in fens, which are so slightly removed now and then, that the lower part remains, sometimes for twenty years together, and in it a collection of spittle, vomit, urine of dogs and men, beer, scraps of fish, and other filthiness not to be named. Hence upon change of weather a vapour is exhaled very pernicious in my opinion to the human body.

(*Letter to Dr. Francis.* Erasmus. 1536.)

INFANT MORTALITY ON ST. KILDA, 1833

'Eight out of every ten children,' he says, 'die between the eighth and twelfth days of their existence!' . . . Mr. Maclean expressly states that 'the air of the island is good and the water excellent'; that 'there is no visible defect on the part of Nature'; and that, on the contrary, 'the great, if not the only, cause is the filth amidst which they live, and the noxious effluvia which pervades their houses'. In proof of this, he refers to 'the clergyman, who lives exactly as those around him do in every respect, except as regards the condition of his house, and who has a family of four children, the whole of whom are well and healthy'; whereas, according to the average mortality around him, at least three out of the four would have been dead within the first fortnight. When it is added, that the huts of the natives are small, low-roofed, and without windows, and are used, during winter, as stores for the collection of manure, which is carefully laid out upon the floor and trodden under foot until it accumulates to the depth of several feet, the reader will not hesitate to concur in opinion with Mr. Maclean, and admit that had the clergyman's children been subjected to the same mismanagement as those of the other islanders, the probability is, that not one of them would have survived.

(Report on the State of Large Towns and Populous Districts. 1844.)

The disease from which these infants perished was tetanus or lockjaw, the bacterium of which is constantly to be found in manure. The navel-string of the babies became infected by the manure, and only by good fortune did any survive. Mr. Maclean detected that dirt was the cause of the mortality, but in what manner that operated could not be known till the discovery of the tetanus bacterium in 1884.

SANITATION IN EDINBURGH, 1769

. . . 'But although I have made these excuses for the nastiness of this place, yet cannot the fact be denied. In a morning earlier than seven o'clock, before the ——— are swept away from the doors, it stinks intolerably; for after ten at night you run a great risk if you walk the streets of having——— thrown upon your head; and it sounds very oddly in the ear of a stranger to hear all passers-by cry out, as loud as to be heard to the uppermost storeys of the houses, which are generally six or seven high in the front of the High Street, "Houd yare hoand"; that is, hold your hand, and throw not till I am past.'

(*A Tour through Great Britain in* 1769.)

HOUSING OF THE WORKERS IN 1840

"In the year 1836," says one of the medical officers of the West Derby Union, "I attended a family of thirteen—twelve of whom had typhus fever, without a bed in the *cellar*, without straw or timber shavings—frequent substitutes. They lay on the floor, and so crowded that I could scarcely pass between them. In another house I attended fourteen patients: there were only two beds in the house. All the patients lay on the boards, and during their illness never had their clothes off. I met with many cases in similar conditions; yet amidst the greatest destitution and want of domestic comfort, I have never heard, during the course of twelve years' practice, a complaint of inconvenient accommodation." Now this want of complaint under such circumstances appears to me to constitute a very melancholy part of this condition. It shows that physical wretchedness has done its worst on the human sufferer, for it has destroyed his mind. . . . I have sometimes checked myself in the wish that men of high station and authority would visit these abodes of their less fortunate fellow-creatures, and witness with their own eyes the scenes presented there; for I have thought that the same end might be answered in a way less disagreeable to them. They have only to visit the Zoological Gardens, and to observe the state of society in that large room which is appropriated to a particular class of animals, where every want is relieved, and every appetite and passion gratified in full view of the whole community. In the filthy and crowded streets in our large towns and cities you see human faces retro-grading, sinking down to the level of these brute tribes, and you find manners appropriate to the degradation. . . . In many courts there is only one supply of water for all its inhabitants and it occupies a good deal of time to procure it and carry it back to the different rooms, where it soon becomes covered with black scum. There is generally a filthy accumulation on the surface of the water in the water-butts. In some courts there is no supply of water; such is the case in Ireland Court and Lusigneas-buildings, Red Lion-street, Spitalfields, which I visited to-day. One woman informed me that her husband lay dead, and that she could not obtain water without the greatest difficulty to wash his "rags". I went into her room, and found her husband lying dead in a coffin; the room was small, dark, and dirty, and occupied by six children, in addition to the father and mother. Another female represented the place to be "stinking alive" for the want of water. In the neighbourhood of Field-lane some persons have not even cesspools or privies; all their excrements are thrown into a little back yard, where they are allowed to accumulate for months together; others have a cesspool, but it is not provided with a drain, so that the excrements run into courts or streets, where they remain until a shower of rain washes them into the gutter. These are the places we are called upon most frequently to visit. . . .

Back Queen-street (south of Queen-street) is approached by several lobbies leading from Queen-street. A visitor, on entering the former, finds himself facing a row of privies of more than 100 yards long. The doors of the privies are about six feet from the house doors opposite; and the space between one privy and another is filled up with all imaginable and unimaginable filth; so that the street consists of a passage little more than six feet wide, with dwelling houses on one side, and a continuous range of necessaries, pigsties, middens, heaps of ashes, etc., etc.; on the other, with a filthy and sluggish surface drain running along one side. The doors opening into this street are, in some cases, the back doors of the Queen-street houses; but 12 houses have their *only* outlets—doors and windows—upon this disgusting and pestiferous passage. According to the returns for the year ending June 30, 1841, the deaths in Queen-street, Back Queen-street, and Queen-street Court, were for that year 36, or 1 death to 14.6 persons.

(Modern death-rate, about 1 per 100.)

Some time ago I visited a poor woman in distress, the wife of a labouring man. She had been confined only a few days, and herself and infant were lying on straw in a vault through the outer cellar, with a clay floor, impervious to water. There was no light nor ventilation in it, and the air was dreadful. I had to walk on bricks across the floor to reach her bedside, as the floor itself was flooded with stagnant water. This is by no means an extraordinary case, for I have witnessed scenes equally wretched; and it is only necessary to go into Crosby-street, Freemasons'-row, and many cross streets out of Vauxhall-road, to find hordes of poor creatures living in cellars, which are almost as bad and offensive as charnel-houses. In Freemasons'-row, about two years ago, a court of houses, the floors of which were below the public street, and the area of the whole court, was a floating mass of putrefied animal and vegetable matter, so dreadfully offensive that I was obliged to make a precipitate retreat. Yet the whole of the houses were inhabited!

"Frying-pan alley was a very famous alley in Holborn. . . . It was very narrow, the only necessary accommodation being at the end. In the first house that I turned into there was a single room; the window was very small and the light came through the door. I saw a young woman there, and I asked her if she had been there some little time. 'Yes,' she said, her husband went out to work, and was obliged to come there to be near his work. She said, 'I am miserable.' 'What is it?' I asked. 'Look there,' said she, 'at that great hole; the landlord will not mend it. I have every night to sit and watch, or my husband sits up to watch, because that hole is over a common sewer and the rats come up, sometimes twenty at a time, and if we did not watch for them they would eat the baby up.' "

THE NEED FOR LEGISLATION

You say that you have generally found employers willing to do anything which does not entail great expense: is this very generally the case?—Very generally; but I ought to add, that when the greatest willingness has been shown, very slight obstacles have sufficed to prevent these good intentions from being carried into effect. This happened in the case of a poor water-gilder, suffering from trembling palsy, caused by frequent exposure to the fumes of mercury. I suggested to his employer a very simple plan of getting rid of these poisonous fumes, and he promised willingly and gratefully to adopt it. I called after a few days and found that he contemplated some alteration in his premises in two or three months' time, that he had thought of a plan that seemed to him preferable, and would then adopt it. Though evidently a humane and intelligent man, he seemed to think as little of this delay as if the health and life of a fellow-creature were not in question. Thus it is with all classes. They form a low estimate of the value of life and health.

(From the Report cited on p. 201.)

CHARLES DARWIN GIVES UP HIS INTENDED MEDICAL CAREER
(1825)

I attended on two occasions the operating theatre in the hospital at Edinburgh and saw two very bad operations, one on a child, but I rushed away before they were completed. Nor did I ever attend again, for hardly any inducement could have been strong enough to make me do so; this being long before the blessed days of chloroform. The two cases fairly haunted me for many a long year.

(From a private letter.)

SEPSIS IN HOSPITALS

There is no hospital, however small, airy or well regulated, where this epidemic ulcer is not to be found at times; and then no operation dare be performed; every cure stands still, every wound becomes a sore, and every sore is apt to run into a gangrene: but in great hospitals especially it prevails at all times, and is a real gangrene; it has been named the hospital gangrene, and such were its ravages in the Hôtel Dieu of Paris (that great storehouse of corruption and disease), that surgeons did not care to call it by its true name, they called it the rottenness, foulness, sloughing of the sore! The word, hospital gangrene, they durst not pronounce! for it sounded like a death-bell; and the hearing of that ominous word, the patients gave themselves up for lost.

(*Principles of Surgery* 1801-7. John Bell.)

PASTEUR PROVES THAT FERMENTATION AND PUTREFACTION ARE BROUGHT ABOUT NOT BY THE AIR BUT BY PARTICLES SUSPENDED THEREIN

I believe I have rigorously established in the preceding chapters that the organised productions of infusions that have previously been heated have no other origin than the solid particles which the air always carries and constantly lets fall on all objects. If there can still remain in the mind of the reader even the least doubt in this regard, it will be removed by the experiments of which I am about to speak.

I place in a round glass flask one of the following liquids, all of which are very easily altered by contact with ordinary air, watery extract of yeast, sweetened watery extract of yeast, urine, beetroot-juice, watery extract of pepper; then I draw out, at the blowpipe, the neck of the flask so as to give it various curatures as is indicated in Pl. I, Fig. 25, A, B, C, D (Fig. 45 of this work). I then bring the liquid to the boil for several minutes until the steam issues freely from the open end of the drawn-out neck, using no other precaution. I then let the flask cool. It is a remarkable fact, calculated to astonish anybody who is used to the delicacy of experiments relative to so-called spontaneous generations, that the liquid in the flask remains indefinitely without alteration. One can handle the flask without apprehension, move it from one place to another, allow it to undergo all the temperature-changes of the seasons, and the liquid in it shows not the slightest alteration and keeps its smell and taste: it is an excellent *conserve d'Appert*.* There will be no change in its nature other than, in certain cases, a direct oxidation, purely chemical, of the matter. But we have seen by the analyses which I have published in this memoir how this action of oxygen is limited, and that in no case are any organised productions developed in the liquids.

It appears that the ordinary air, returning with force at the first moment, must reach the flask quite unaltered. That is

* Appert, a Frenchman, was the first to preserve food by heat in the fashion of the modern canning and bottling.

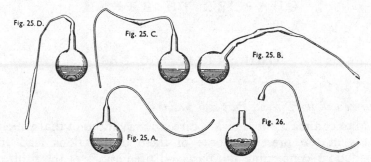

FIG. 45.—The flasks used by Pasteur in his crucial experiments
on spontaneous generation.

true, but it encounters a liquid that is still near its boiling-
point. The re-entry of the air thereafter occurs more slowly,
and by the time that the liquid is so far cooled as no longer to
be able to deprive germs of their vitality, the re-entry of the air
becomes so slow that it leaves behind in the damp curves of
the neck, all the dust particles capable of acting on the infusions
and bringing about the production of organism. At least, I see
no other possible explanation of these curious experiments. If,
after one or many months' stay in the incubator, the neck of the
flask be detached by a nick with a file, without touching the
flask in any other way (*Pl. I, Fig* 26, *V, Fig.* 45 *of this work*),
after twenty-four, thirty-six or forty-eight hours, moulds and
infusoria begin to show themselves absolutely in the ordinary
way or just as if one had sown dust particles from the air in the
flask. . . .

LISTER REPORTS HIS SUCCESS (1867)

Since the antiseptic treatment has been brought into full
operation, and wounds and abscesses no longer poison the atmosphere with putrid exhalations, my wards, though in other
respects under precisely the same circumstances as before, have
completely changed their character; so that during the last nine
months not a single instance of pyæmia, hospital gangrene or
erysipelas has occurred in them.

(Note that Lister at this date believes that the main channel of
infection of wounds are 'exhalations' mingling with the air.)

CHAPTER THIRTEEN

Man Learns the Past

Science and the Present, Past and Future

Man can make more or less direct observations of that which exists in the present. Some of these observations lead to deductions concerning the past and predictions of the future. The past and future must always be known with less certainty than the present; nevertheless where evidence is plentiful we can have much confidence in our history or prediction—but where it is scanty, we must accept the results of our reasoning with due reserve.

Traditional views of the past

An account of the early history of the world is given in the Book of Genesis. From the time of St. Augustine (fifth century) at least, it has been recognised that this is not necessarily to be read by the Christian as literal scientific history, but that it is told in such a way as to be understood by its ancient hearers, yet remaining true at all times for all men who had the wits to discover its meaning.

This enlightened view was not however general in the eighteenth and early nineteenth centuries, and the majority at that time considered that the world, with its plants, animals and men, such as exist to-day, was created in six successive days of twenty-four hours, about the year 4000 B.C.

Fossils

The most obvious difficulty in harmonising the world as we see it to-day with the literal interpretation of the Mosaic view of Creation is the character of fossil remains of living creatures. How did oyster shells get into the limestone of the Alps? Had the Alps been under the sea at some time since their creation? If so, how was it that no such changes as would raise the Alps in so short a period were still to be observed? That astonishing genius Leonardo da Vinci solved the problem in outline in his notebooks written between 1482 and 1518, but unfortunately for science not published for centuries after his death. Leonardo brings strong evidence that fossils were once living and that they were covered up by mud which was changed into rock. The sea-bed rose to form the dry land as a result of the shifting of the centre of gravity of the earth through the erosion and lightening of the land and the deposition of mud in the sea. But Leonardo was almost alone in these advanced views, and the favourite theory was that a *formative power*, which was thought to cause all living things, and crystals likewise, to assume their characteristic shapes, could act on the rocks and produce there the forms of shells, leaves, bones, etc., which forms had, however, never been alive. This view was generally, though not invariably, accepted up till the beginning of the eighteenth century. When, however, the first naturalists began to classify fossils, as they classified living plants and animals, it became obvious that these fossils were remains of actual creatures that had once been alive and of species that were no longer to be found on earth. How had these species become extinct? The Biblical flood was generally invoked both to destroy these creatures and to carry them to the tops of the mountains, and when it became clear that different species of animals had become extinct at different periods, a series of floods or catastrophes was posited. But what were the naturalists to think of the animals, living in modern times, whose remains were not to be found in the deeper rocks? Had these animals come into being after the general creation? All species, it was supposed, had been created by the direct fiat of the Deity: and a usual interpretation was to suppose the six days' of creation to refer to six great geological periods. But as research went on, it did not seem that there were precisely six or any other such number of geological periods, but a continuous process of extinction and new creation. In 1858, just before Darwin's theories were published, the favoured view was that a creative force in Nature, emanating originally from God, had created new species at intervals throughout the ages. It was supposed only by very few that one species could change into another, for the excellent reason that no evidence of such changes had been observed. Although no clear account of the appearance and disappearance of fossil forms could yet be given, they were exceedingly carefully studied by such men as Cuvier and Owen; and palæontology, the study of ancient forms of life, was well advanced by the middle of the nineteenth century.

Geological time

Between 1790 and 1820 a number of great geologists discovered how the various strata had been laid down. J. G. Werner made it clear that successive strata or layers of rock succeeded each other in time, the deepest being the oldest. William Smith showed how each stratum could be identified by the particular fossil species it contained, and by such means geological maps began to be prepared. James Hutton in 1795 showed how the effects of weathering and the mechanical action of water disintegrated the rocks and led to the deposition of such strata as clays, shales, slates, sandstones, etc. He made it clear that the deposition of these rocks in such gigantic thicknesses as were now revealed by geological researches must have required incalculable ages. The earth was obviously very much more than six thousand years old, and the 'days' before the creation of man were to be thought of in terms of thousands, even millions of years. Physics had its contribution to make to this problem; the earth had once been a molten globe, as appeared from its oblate spheroidal shape: how long had it needed to cool to its present temperature? There was hardly sufficient data to answer this, for the cooling of the earth was influenced by the heat of the sun and this itself raised further problems (pp. 220, 242–45).

Views of the Origin of Species before Charles Darwin

The species of animals and plants are obviously not unrelated. Thus there are several hundred species of plants of the onion tribe, several hundred species of bee, several dozen species of mice. A scientific explanation of this fact was required, and a possible one was that all the similar species had developed from a single common stock, as all the domestic varieties of the dog were taken to have developed from one (or a very few) species of wild dog. The view that all animal species had developed from a single type of primitive organism was advanced in 1794 by Erasmus Darwin, Charles Darwin's grandfather, but it was considered a fanciful and eccentric hypothesis (pp. 221–24). Lamarck (1809) suggested a similar theory and explained the alteration of species as being due to the effort of the species to adapt itself to its mode of living. Thus, on the Lamarckian theory, the long neck of the giraffe was produced by the *efforts* of an antelope-like ancestor to stretch up to higher and higher branches of the trees on the leaves of which it fed. This theory was not given serious consideration until the work of Charles Darwin revived interest in these problems.

Charles Darwin

Charles Darwin was born in 1809. He first intended to enter the medical profession, then the clergy, but after taking his degree at Cambridge almost at once joined the surveying expedition of the "Beagle" as naturalist. During this voyage he was struck by the extremely odd flora and fauna of oceanic islands, such as the Mauritius and Galapagos Islands, where perhaps three-quarters of the species might be peculiar to one isolated island. If these species had *evolved* by descent from others, then the long-isolated population of these islands would be expected to be different from those of each other and the mainland, for external conditions on the island would differ from those of the other islands or the mainland, and evolution would not take the same course; but on a doctrine of the separate creation of each species there seemed to be no reason why isolated islands should be provided with highly aberrant species, e.g. the dodo, the giant tortoise, etc. He was impressed by the fact that the recent fossils of a region were related to, but not identical with, the living animals of that region. The alterations that have come about in domesticated animals and plants likewise struck him forcibly. Man changes these species by selecting the animals or plants from which he wishes to breed, but what agency could be bringing about these changes under natural conditions? Some light was thrown on this by the theories of T. R. Malthus, the economist. He laid down the principle (1798) that the population increases in geometrical ratio but subsistence at best only in arithmetical ratio, so that the increase of population must, if unrestrained, be checked by want. Darwin realised that individual animals of the same species vary slightly, and all animals are producing vastly more offspring than survive to maturity: those that do survive will not be average specimens of the race, but will be those individuals best fitted to find food, escape their enemies, etc. The process by which the fit survive in preference to the less fit he termed *natural selection*. Darwin supposed that these variations were inherited, and that the survivors would therefore pass on their 'fitness' to their descendants, who would therefore be 'fitter' specimens than the average of the previous generations; the whole species will therefore change in the direction of 'fitness', i.e. the possession of those qualities that enabled them to survive and produce offspring. Darwin developed the notion very slowly and cautiously, and in 1858 Alfred Russell Wallace reached the same conclusions. They did not quarrel about priority, but published a joint contribution. Then in 1859 Darwin published his *Origin of Species by Means of Natural Selection*. Darwin claimed that the propositions put forward in this paragraph and given in his own words on pp. 226–28 served to explain the development of the complex organs and instincts of animals by an accumulation of innumerable slight variations, each 'good' for the individual. He brings strong evidence that species have in fact evolved, but he does not clearly prove that natural selection is the means of bringing about that evolution. He could not show that variations sufficiently large to be of use in bringing about survival did occur or were inheritable: nor was it at all clear how a chance variation in an individual could fail to be swamped by interbreeding with other individuals in whom such variations were not present. Nevertheless Darwin's theory made the whole world of living organisms into an *intelligible pattern*, a genealogical tree rooted in the primitive protozoon and extending to every present living type. It explained the fact that, e.g. all mammals, or insects, or worms, are variations of a single type; that the wing of a bat, the flipper of a seal, the hand of a man, the hoof of a horse, though used for such diverse ends, are yet of the same pattern, a fact which was inexplicable on any theory other than that of the lineal descent of all these creatures from a common ancestor; it explained likewise such strange phenomena as the existence of useless leg-bones within the body of a whale, and of fish-like gill-arches in the human foetus.

The Study of Heredity

The Darwinian case then depended on the "inheritance of the small variations", and this was in fact unproven. The improvement in the theory of evolution came from a study of heredity. Weissmann (1882) attacked the doctrine of inheritance of acquired characters. Darwin had tentatively regarded inheritance as *pangenesis*, 'elements' from the whole body uniting in the germ-cell: on this hypothesis any change in the body of an individual might alter its germ-cells and so be inherited. Weissmann regarded the germ-plasm as the permanent transmissible unit and the individual as an impermanent outgrowth that perishes. Changes in the individual could not, he held, affect the germ-plasm.

The next piece of important work (earlier in time but later in effect) was that of Mendel, abbot of the Augustinian monastery of Brünn in Moravia. His experiments were carried out in 1866–9, and attracted little or no attention. Mendel studied the inheritance of various characters in the culinary pea. He crossed round-seeded and crumpled-seeded peas; green-pod and yellow-pod; tall and dwarf; and he found that there were hereditary factors, such as 'yellow-seededness', which were inherited as a whole and not in varying degrees. He postulated that corresponding to every inherited character there are determinants or factors (now called genes) in the gametes. Each individual has a double set of factors, and the character of the offspring will depend on whether it has none, one or two of these. Mendel's work was not given the attention until the year 1899 when Bateson rediscovered these phenomena, only to find himself partly anticipated. From 1910 it became clear that the hereditary factors or genes were localised portions of the chromosomes of the cells, and it followed that evolution must involve changes in these genes. Such changes have been brought about experimentally, e.g. by X-ray treatment, and it is quite clear that alterations in the chromosomes (and therefore genes) produce variations that breed true.

It has also been made clear that changes in the chromosomes do occur without any known cause and give rise to variations in individuals, which may be great or small, advantageous or otherwise (e.g. hæmophilia). The present view, then, is that mutations in the germ-plasm do occur, by the operation of some unknown causes. These mutations are inherited and *may* be carried by individuals who do not display their characters. Thus a new mutation occurring naturally is not swamped by crossing with those that do not possess it, but rather spreads to their progeny. If the results of the mutation are disadvantageous, the individuals that carry it will be removed by the Darwinian process of natural selection: if it is advantageous, those having it will be selected. What is not clear is (*a*) how a separate species not breeding with other species is formed, (*b*) what agency produces the large number of mutations in the same direction which would be needed to produce a feather from a scale, or man's brain from an ape's. In absence of knowledge as to the mechanism of mutation it is unwise to be dogmatic.

The Control of Life

In the Middle Ages and earlier times man was aware of a general plan by which the mineral world was the support of the kingdom of plants, which in turn served to nourish the kingdom of animals, whose function was to feed and serve man, its culmination. Whether or no modern man believes this to be a deliberate design, he is rapidly carrying it to its ultimate limit, at which the surface of the globe will be a scientific and highly productive farm. Ever wider areas are being brought under cultivation. Air, water and the rocks are being converted into fertilisers, to nourish plants; these in turn are supporting huge populations of animals, to be killed early in their highly artificial lives to feed the increasing population of men. This increase of production has had its disasters. The invasion of Australia by the rabbit and prickly-pear, the creation of 'dustbowl' conditions in the U.S.A. are the result of man's imperfect knowledge of biology. Ecology, the study of the balance that prevails between species and which man upsets, makes it possible to restore such catastrophes and prevent others.

But has man the right to destroy the wild, beautiful and cruel life of Nature, to exterminate some species and to convert others into passive meat-producers? Science has no answer, for the word 'right' is not found in its vocabulary.

The Idea of Progress

The idea of the evolution of species made a great stir in the world. At first there was intense opposition, due to the apparent conflict of the theory with the Book of Genesis—especially with regard to the descent of man from an apelike ancestry: this gradually died down as men became accustomed to a freer interpretation of the Holy Scriptures, and distinguished more carefully the descent of the body and the origin of the soul. On the other hand, the theory was welcomed in many quarters as contributory to the notion of Progress which was very dear to the Victorians.

This notion of Progress had not been an effective one before the sixteenth century. The men of earlier times did not look on the future good of the human race in general as the end they sought to attain. It was quite clear to them that each man had an individual destiny and that his purpose was to serve God and by serving Him to attain everlasting blessedness. It was clear to them that the great men of the past were as great as or greater than those of their own age, and if the habits of life of the fifteenth century were more polished than those of the eighth, so much the worse for the men of the fifteenth century who had turned their minds from God to run after country-houses, with carpets on the floor, glass in the windows, and silks and satins from the East. In the sixteenth century this attitude persisted: the Elizabethans might pride themselves on a superiority in manners to a rude ancestry, but they counted themselves greatly inferior to the worthies of Greece and Rome. When, however, the ancient science came to be cried down by Bacon, Galileo, Descartes and others, and the new scientific discoveries were seen to be something useful which no previous age had possessed, there came into being the notion that *man was progressing*. The power of man to provide himself with what he wanted was obviously growing, and his knowledge was rapidly increasing in volume and certainty. Material progress of that type there certainly was; but more than this was claimed by the exponents of progress, notably in France. An Age of Reason was coming in. Men were no longer to be swayed by superstition; they were to obey the natural promptings of good sense and reason, whereupon the iniquities of war, persecution, and crime, would disappear. The French revolution was felt to be the beginning of that age, and there is no need to tell how the subsequent course of events threw doubt on the power of reason to control man's passions and their expression in action.

Parallel with this movement in France was a different one in England, where utilitarianism came into fashion. Reason again was the mistress, but instead of being manifest in pure science and philosophical ethics as in France, it was of an industrial and practical turn. The greatest good of the greatest number was preached. Wealth was to increase through production and cheapening of goods. If a few tens of thousands of workers starved—well, it was just an incident in progress towards prosperity. The cotton-mills, steam-engines, gas-light, steamships, balloons—the swelling tide of new discovery gave the English a powerful sensation of progress towards a blessed state of prosperity, education and peace; and this is true even of the 'forties when the country was suffering extremes of famine and disease, which fell, however, chiefly upon the unexpressive poor. Yet even the oppressed, the Chartists for example, believed a golden age was coming; thus the works of Karl Marx and Engels, who were young at this period, show a belief that man would progress to a perfected individual and social way of life. The prosperity which followed on the Great Exhibition of 1851 renewed the certainty of progress, and it was in this atmosphere that Darwin concocted and finally launched his theory of Evolution. The idea of continual change for the better was there already; for any voluntary change is thought by self-deluded man to be progress inasmuch as it is the realisation of what he thinks desirable. Here, then, was Darwin, presenting the whole history of the world in a form that could be conceived of as a *continuous* perfecting of living creatures, the crown of all being Man, who was presumably still evolving to a greater perfection. Laplace's nebular theory (pp. 240–42) had given a picture of formless gas evolving to the complex solar system: this with Darwin's theory of Evolution painted on a huge canvas the scientific view of the history of the world as a progress from formless gas to Queen Victoria—and so onward to the super-man of the future. The whole world seemed to be undergoing a ceaseless process of change, transforming the simple and mindless into the complex and intelligent. Small wonder that when doubt was being cast on the foundations of Christianity, this vast plan was made the foundation of a religious philosophy or at least a system of ethics. The influential writer in this field was Herbert Spencer, whose writings covered the half-century 1850–1900. He was the inventor of the word *Evolution* and of the phrase "survival of the fittest". According to Spencer the universe oscillated between Evolution towards the higher (which was the more complex) and Dissolution which was the opposite process. Evolutionary ethics would seem to indicate that everyone should take steps to ensure his own survival: this Spencer humanised by the argument that living beings must (on pain of extinction) take pleasure in actions that conduce to their survival. Good conduct, therefore, is that which is conducive to the preservation of a pleasurable life in a society so adjusted that each attains his happiness without impeding that of others. All this is somewhat remote from the biological theory of the evolution of species, but the fact remains that the scientific Victorians felt themselves to be in a state of very rapid progress to something higher, without being able to define clearly what was higher and what was not. When the turn of the century came the idea of a Law of Progress seemed more doubtful, and the war of 1914–18 gave it a blow from which it had not staggered to its feet before the events of the 'thirties and 'forties gave it what may be its death-blow.

But popular beliefs are long a-dying, and this generous faith in progress is still abroad, especially in the U.S.A. and U.S.S.R., which have been undergoing a process of expansion and increasing prosperity very like that of Victorian England. It is as well therefore to analyse the idea of progress a little further, and especially to distinguish between the progress of species as a result of biological evolution and the progress of human institutions as a result of man's thought and effort.

Some Victorians took the Darwinian point of view and supposed that *man himself* was progressing. The highest type of man, who wore a collar and carried an umbrella, lived a healthier life and had families of whom more survived, than the lower and obviously undesirable, dirty, and immoral poor. Therefore the former would tend to survive and to transmit to his offspring the qualities that had enabled him to bring them up so well. All that looks very different now. In the first place we are not at all sure what qualities we should regard as those whose increase would constitute progress in man. But if they are intelligence, thrift, imagination, etc., it is quite clear that in fact the conspicuous possessors of these qualities are bringing up fewer children than anyone else, and so are presumably on the road to extinction. Fortunately, however, we do not regard the matter as so simple; and we have little doubt that any social section, except perhaps the submerged (which has a high proportion of subnormal intellects), will yield very much the same heredity.

Secondly, we realise that while the breeding of new human varieties, i.e. reshuffling of hereditary factors, is possible in theory (though excessively difficult to bring about in practice) true evolution is exceedingly slow; and we have no evidence that man's central nervous system has altered by a trace in fifty thousand years. We do not know where *Homo Sapiens* came from, and we have no reason to suppose he is changing more or more rapidly than the cat or the candytuft, nor do we know what chain of events could lead to the production of a *Homo Sapientior*. It is true that Evolution, broadly viewed, shows a progress from "amœba" to man, but it equally shows a progress from amœba to shark or amœba to hook-worm, to name three of the most flourishing organisms. There is no reason to put man at the top of the evolutionary tree unless we make the arbitrary choice of brain or mind as the topmost of attributes. On the showing of the biologists, brain or mind has been evolved, because it *paid* in survival value: if brain or mind operates against survival or fertility, the evolutionist must suppose that it will die out in a period of geological time.

It follows then that if we are to make evolution our criterion of progress we are not enlisting under the standard of mind but of multiplication and self-preservation, for that is the only way it can lead us.

That there is a progress associated with modern man is perfectly clear. It is not a progress in the native faculties of the babies born (which evolution would bring about if it operated) but a progress in the variety and extent of knowledge and power open to them as they develop and mature. Man remains the same, but he continually develops the means of equipping his brain and enabling his ideas to have practical effect. Science is *cumulative*; everything that each age discovers remains for all ages, and consequently man is receiving an ever greater birth-right of power, but little or no increased judgment in its use. The spectacle of the world to-day is the clearest demonstration of this danger. The power of the man of 1990 will exceed that of 1940, to a greater extent than that of 1940 exceeded that of 1890—before the automobile, electrical supply and aeroplane: history, moreover, gives us reason to suspect that the ability of the unscrupulous to sway man's passions will be no whit abated.

If civilisation is to survive it must discover, adopt, and enforce, an absolute ethical standard. Such a standard is to be found in religion, and it has not been found elsewhere. It is true that at no period of human history has religion been able to restrain more than a part of the world's wickedness, but it is conceivable that if some of the efforts we have devoted to persuading men to buy luxuries were devoted to trying to persuade them to behave as Christians, a public opinion might be built up capable of restraining the excesses that now threaten to destroy us. Man is behaving like a cunning brute because he has convinced himself that science has proved that he is no more than that. Let the nature of scientific reasoning and evidence be understood; and its conclusions be applied to the material realm to which alone they are relevant; then the peculiar eminence of man and the excellence of the good of which he is capable will be once more apparent. The way will then be open for a true consideration of man as organism, as mind, and as soul, from which true consideration will emerge standards of conduct, true, wise, and in the highest sense human, because divine.

Examples on Evolutionary Theories

JAMES HUTTON PROVES THE EARTH'S ANTIQUITY

To sum up the argument, we are certain, that all the coasts of the present continents are wasted by the sea, and constantly wearing away upon the whole; but this operation is so extremely slow, that we cannot find a measure of the quantity in order to form an estimate: therefore, the present continents of the earth, which we consider as in a state of perfection, would, in the natural operations of the globe, require a time indefinite for their destruction.

But, in order to produce the present continents* the destruction of a former vegetable† world was necessary; consequently the production of our present continents must have required a time which is indefinite. In like manner, if the former continents were of the same nature as the present, it must have required another space of time, which also is indefinite, before they had come to their perfection as a vegetable world. . . .

The result, therefore, of this physical inquiry is, that we find no vestige of a beginning,‡—no prospect of an end.

(*Theory of the Earth*. James Hutton. 1795.)

* Which are for the most part of sedimentary rock.

† Vegetable—used in obsolete sense: capable of supporting life.

‡ He does not disprove such a beginning but finds no geological evidence of it. How far is this sound?

ERASMUS DARWIN ON THE RELATION OF SPECIES

[It is to be noticed that Erasmus Darwin has advanced evidence for the fact of evolution, but no theory of its mechanism.]

"Secondly, when we think over the great changes introduced into various animals by artificial or accidental cultivation, as in horses, which we have exercised for the different purposes of strength or swiftness, in carrying burthens or in running races; or in dogs, which have been cultivated for strength and courage, as the bull-dog; or for acuteness of his sense of smell, as the hound and spaniel; or for the swiftness of his foot as the greyhound; or for his swimming in the water, or for drawing snow-sledges, the rough-haired dogs of the north; . . . and add to these the great changes of shape and colour, which we daily see produced in smaller animals from our domestication of them, as rabbits, or pidgeons; or from the difference of climates, and even of seasons; thus the sheep of warm climates are covered with hair instead of wool; and the hares and partridges of the latitudes which are long buried in snow, become white during the winter months; . . .

"Fourthly, when we revolve in our minds the great similarity of structure which obtains in all the warm-blooded animals, as well as quadrupeds, birds and amphibious animals, as in mankind; from the mouse and bat to the elephant and whale; one is led to conclude, that they have alike been produced from a similar living filament.* In some this filament in its advance to maturity has acquired hands and fingers, with a fine sense of touch, as in mankind. In others it has acquired claws or talons . . . in others toes with an intervening web or membrane . . . in others it has acquired cloven hoofs . . . and whole hoofs in others . . . while in the bird kind this original living filament has put forth wings instead of arms or legs, and feathers instead of hair. In some it has protruded horns on the forehead instead of teeth in the fore part of the upper jaw; in others tushes instead of horns; and in others beaks instead of either. And all this exactly is daily seen in the transmutations of the

* Primitive organism: supposed by Erasmus Darwin to be analogous to a spermatozoon.

tadpole, which acquires legs and lungs when he wants them; and loses his tail when it is no longer of service to him.

"Fifthly, from their first rudiment, or primordium, to the termination of their lives, all animals undergo perpetual transformations, which are in part produced by their own exertions in consequence of their desires and aversions, of their pleasures and pains, or of irritations, or of associations; and many of these acquired forms or propensities are transmitted to their posterity.

"As air and water are supplied to animals in sufficient profusion, the three great objects of desire, which have changed the forms of many animals by their exertions to gratify them, are those of lust, hunger and security. A great want of one part of the animal world had consisted in the desire of the exclusive possession of the females; and these have acquired weapons to combat each other for this purpose, as the very thick, shield-like, horny skin on the shoulder of the boar is a defence against only animals of his own species, who strike obliquely upwards, nor are his tushes for other purposes, except to defend himself, as he is not naturally a carnivorous animal. So the horns of the stag are sharp to offend his adversary, but are branched for the purpose of parrying or receiving the thrusts of horns similar to his own, and have, therefore, been formed for the purpose of combating other stags for the exclusive possession of the females; who are observed, like the ladies in the time of chivalry, to attend the car of the victor.

"The birds which do not carry food to their young, and do not therefore marry, are armed with spurs for the purpose of fighting for the exclusive possession of the females, as cocks and quails. It is certain that these weapons are not provided for their defence against other adversaries, because the females of these species are without this armour. *The final cause of this contest amongst the males seems to be, that the strongest and most active animal should propagate the species, which should thence become improved.*

"Another great want consists in the means of procuring food, which has diversified the forms of all species of animals. Thus the nose of the swine had become hard for the purpose of turning up the soil in search of insects and of roots. The trunk of the

elephant is an elongation of the nose for the purpose of pulling
down the branches of trees for his food, and for taking up water
without bending his knees. Beasts of prey have acquired strong
jaws or talons. Cattle have acquired a rough tongue and a
rough palate to pull off the blades of grass. . . . Some birds have
acquired harder beaks to crack nuts, as the parrot. Others
have acquired beaks adapted to break the harder seeds, as
sparrows. Others for the softer seeds of flowers, or the buds
of trees, as the finches. Other birds have acquired long beaks
to penetrate the moister soils in search of insects or roots, as
woodcocks, and others broad ones to filtrate the water of lakes,
and to retain aquatic insects. *All which seem to have been gradually
produced during many generations by the perpetual endeavour of the
creatures to supply the want of food, and to have been delivered to their
posterity with constant improvement of them for the purpose required.*

"The third great want among animals is that of security,
which seems much to have diversified the forms of their bodies
and the colour of them; these consist in the means of escaping
other animals more powerful than themselves. Hence some
animals have acquired wings instead of legs, as the smaller
birds, for the purpose of escape; others great length of fin or
of membrane, as the flying fish and the bat. Others great
swiftness of foot as the hare. Others have acquired hard or
armed shells, as the tortoise and the echinus marinus.

"The contrivances for the purposes of security extend even
to vegetables, as is seen in the wonderful and various means
of their concealing or defending their honey from insects, and
their seeds from birds. On the other hand, swiftness of wing
has been acquired by hawks and swallows to pursue their prey;
and a proboscis of admirable structure has been acquired by
the bee, the moth, and the humming bird for the purpose of
plundering the nectaries of flowers. All which seems to have
been formed by the original living filament, excited into action
by the necessities of the creatures, which possess them, and on
which their existence depends.

"From thus meditating on the great similarity of the structure
of the warm-blooded animals, and at the same time of the great
changes they undergo both before and after their nativity; and

by considering in how minute a portion of time many of the changes of animals above described have been produced; would it be too bold to imagine, that in the great length of time, since the earth began to exist, perhaps millions of ages before the commencement of the history of mankind, would it be too bold to imagine, that all warm-blooded animals have arisen from one living filament which the GREAT FIRST CAUSE endued with animality, with the power of acquiring new parts, attended with new propensities, directed by irritations, sensations, volitions, and associations; and thus possessing the faculty of continuing to improve by its own inherent activity, and of delivering down those improvements by generation to its posterity, world without end!"

(*Zoonomia*. Erasmus Darwin, 1794.)

LAMARCK'S 'LAWS'

[Lamarck supposed that species underwent change and that all were descended from a common ancestor. He regarded the kind and degree of *use* of the organs of the animal as the cause of its modification.]

First Law

In every animal which has not passed the term of its development, a more frequent and sustained use of an organ, gradually strengthens that organ, develops it, enlarges it, and gives it a power proportionate to the duration of that use; while the constant lack of use of such an organ, insensibly weakens it, worsens it, progressively diminishes its faculties and ends by causing it to disappear.

Second Law

All that nature has caused individuals to acquire or lose by the influence of the environment in which their race has been placed for a long period and consequently, by the influence of the predominant use of any organ or by that of a constant lack of use thereof: all this is preserved by reproduction for the new individuals which arise therefrom, provided that the modifications acquired are common to both sexes or at least to the individuals which produce the young.

[Note the difference from Darwin's view.]

(*Philosophie Zoologique*. Lamarck. 1809. pp. 234–5. Translated as *Zoological Philosophy*, by H. Elliot. 1914.)

HUGH MILLER ARGUES AGAINST LAMARCK

. . . The setting dog is taught to set: he squats down and points at the game; but the habit is acquired one,—a mere trick of education. What, however, is merely acquired habit in the progenitor, is found to pass into instinct in the descendant; the puppy of the setting-dog squats down and sets *untaught*.* . . .

The woody plant of a warmer climate, when transplanted into a colder frequently exchanges its ligneous stem for a herbaceous one, as if in anticipation of the killing frosts of winter, and, dying to the ground at the close of autumn, shoots up again in spring. . . . But it is easy driving a principle too far. . . . And such, but still more glaring, has been the error of Lamarck. He has argued on this principle of improvement and adaptation, which, carry it as far as we rationally may, still leaves the vegetable a vegetable and the dog a dog, that, in the vast course of ages, inferior have risen into superior natures and lower into higher races; that molluscs and zoophytes have passed into fish and reptiles . . . and that monkeys and apes have been transformed into human creatures, capable of understanding and admiring the theories of Lamarck. . . .

All the animal families have . . . their connecting links and it is chiefly out of these that writers such as Lamarck and Maillet construct their system. They confound gradation with progress. Geoffrey Hudson was a very short man and Goliath of Gath a very tall one, and the gradations of the human stature lie between. But gradation is not progress, and though we find full grown men of five feet, five feet six inches, six feet, and six feet and a half, the fact gives us no earnest whatever that the race is rising in stature. . . .

* It is doubtful whether this is true.

. . . Now it is a geological fact that it is fish of the higher orders that appear first on the stage, and that they are found to occupy exactly the same level during the vast period represented by five succeeding (geological) formations. There is no progression. If fish rose into reptiles it must have been by sudden transformation;—it must have been as though a man who had stood still for half a life-time should bestir himself all at once, and take seven leagues at a stride.

(*The Old Red Sandstone*. Hugh Miller, 1841, pp. pp. 43–5.)

CHARLES DARWIN ON NATURAL SELECTION

If during the long course of ages and under varying conditions of life, organic beings vary at all in the several parts of their organisation, and I think this cannot be disputed; if there be, owing to the high geometrical powers of increase of each species, at some age, season, or year, a severe struggle for life, and this certainly cannot be disputed; then, considering the infinite complexity of the relations of all organic beings to each other and to their conditions of existence, causing an infinite diversity in structure, constitution and habits, to be advantageous to them, I think it would be a most extraordinary fact if no variation ever had occurred useful to each being's own welfare, in the same way as so many variations have occurred useful to man. But if variations useful to any organic being do occur, assuredly individuals thus characterised will have the best chance of being preserved in the struggle for life; and from the strong principle of inheritance they will tend to produce offspring similarly characterised. This principle of preservation, I have called, for the sake of brevity, Natural Selection. Natural selection, on the principle of qualities being inherited at corresponding ages, can modify the egg, seed, or young, as easily as the adult. Amongst many animals, sexual selection will give its aid to ordinary selection, by assuring to the most vigorous and best adapted males the greatest number of offspring. Sexual selection will also give characters useful to the males alone, in their struggles with other males. . . .

. . . The tubes of the corollas of the common red and incarnate clovers (*Trifolium pratense* and *incarnatum*) do not on a hasty glance appear to differ in length; yet the hive bee can easily suck the nectar out of the incarnate clover, but not out of the common red clover, which is visited by bumble-bees alone; so that the whole fields of the red clover offer in vain an abundant supply of precious nectar to the hive-bee. Thus it might be a great advantage to the hive-bee to have a slightly longer or differently constructed proboscis. On the other hand, I have found by experiment that the fertility of clover greatly depends on bees visiting and moving parts of the corolla, so as to push the pollen on to the stigmatic surface. Hence, again, if bumble-bees were to become rare in any country, it might be a great advantage to the red clover to have a shorter or more deeply divided tube to its corolla, so that the hive-bee could visit its flowers. Thus I can understand how a flower and a bee might slowly become, either simultaneously or one after the other, modified and adapted in the most perfect manner to each other, by the continued preservation of individuals presenting mutual and slightly favourable deviations of structure.

. . . It is interesting to contemplate an entangled bank, clothed with many plants of many kinds, with birds singing on the bushes, with various insects flitting about, and with worms crawling through the damp earth, and to reflect that these elaborately constructed forms, so different from each other, and dependent on each other in so complex a manner, have all been produced by laws acting around us. These laws, taken in the largest sense, being Growth with Reproduction: Inheritance which is almost implied by reproduction; Variability from the indirect and direct action of the external conditions of life, and from use and disuse; a Ratio of Increase so high as to lead to a Struggle for Life, and as a consequence to Natural Selection, entailing Divergence of Character and the Extinction of less-improved forms. Thus, from the war of nature, from famine and death, the most exalted object which we are capable of conceiving, namely, the production of the higher animals, directly follows. There is grandeur in this view of life,

with its several powers, having been originally breathed into
a few forms or into one; and that, whilst this planet has gone
cycling on according to the fixed law of gravity, from so simple
a beginning endless forms most beautiful and most wonderful
have been, and are being, evolved.

(*On the Origin of Species.* Charles Darwin. 1858.)

CHAPTER FOURTEEN

Man Discovers the Universe

The likeness of all things

By the beginning of the eighteenth century, educated people had no longer any reason to believe that in the heavenly bodies there was any kind of matter or any forces operating that did not exist upon earth. The evidence for this similarity of the terrestrial and celestial was still slender. Newton had shown in 1687 that a planet or satellite travelled in its orbit in conformity with the same laws as stated the behaviour of an earthly projectile, e.g. a cannon-ball. Galileo, from 1610 onward, and subsequent telescopic observers, saw that the moon and planets looked much as the earth might have been expected to look at the same distance. In absence of evidence to the contrary it was assumed that planets, suns and stars were composed of the same sort of material as the earth. There might, however, have been in the heavenly bodies any number of chemical elements other than those known on earth. Very little attention was paid to this subject in the eighteenth century, and in the nineteenth Comte declared that the composition of

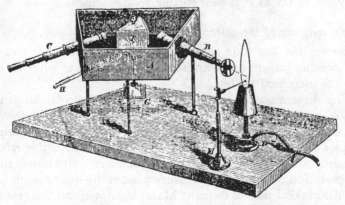

FIG. 46.—The first spectroscope (1859).

the heavenly bodies could never be ascertained by man. Newton had shown that white light could be decomposed by a prism
so as to form a spectrum or band of light in which were visible
all the differently coloured rays that together blended to white
light. Further refinements showed that the spectra of glowing gases were not continuous bands but consisted of bright
separate lines: and that each chemical element in the glowing
gas gave a series of bright lines that were easily recognisable.
Bunsen and Kirchhoff made this discovery in 1859 and in the
same year Kirchhoff studied the spectrum of the sun, and
announced that it contained sodium, iron, magnesium, calcium,
chromium, copper, zinc, barium and nickel. This work was
rapidly extended and we are now aware that the heavenly
bodies consist of the ordinary chemical elements known on
earth: and, while we cannot be sure that there are no others
existing in the interior of the star, which are inaccessible to
our instruments, there is no spectroscopic evidence of their
existence. One element, indeed (helium) was discovered
spectroscopically in the sun before it was known upon earth.

Not only then do the heavenly bodies appear like, and move
like, the earthly, but they are known to be made of the same
stuff. Science, then, derived from the study of things upon
earth, is the key to the understanding of the physical relations
of things, not only upon earth, but throughout the whole
universe as far as it is perceptible to the senses.

The dimensions of the universe

During the last four centuries the amount that we can see
has steadily grown, and our estimate of the size of what we can
see has increased no less.

The dimensions of the solar system began to be known in
the seventeenth century when values were obtained for the
distance of the sun varying from 41 to 136 million miles.
Kepler's third law (p. 116) allows us to deduce the *relative*
distances of the planets and sun from their well-known times
of revolution, so if any one measurement of the solar system (e.g.
the distance of the earth from Mars) is known, all the rest can
be calculated. If Mars is observed from two different observa-

tories (or the same observatory at different stages of the earth's daily rotation) its position relative to the background of fixed stars differs by a few seconds of arc (say about 1/200 of the diameter of the moon). From this small angle its distance can be calculated. Gradual refinements made the distance of the sun known as about 90,000,000 miles at the beginning of the nineteenth century and the diameter of the solar system, before Uranus was added in 1781, and Neptune in 1846, was taken as about 1,500,000,000 miles.

But what of the distance of the stars? The earth's orbit was known to be some 200,000,000 miles in diameter, so on any particular day the earth was that much nearer to some particular constellation than it was six months before. The nearer you are to a thing the bigger it looks (the greater the angle it subtends) yet no change of the kind was visible in the stars. It was clear, then, that the constellations were so far away that an approach of 200,000,000 miles did not appreciably alter the angle they subtended. This seemed at first absurd, then marvellous, to the seventeenth-century astronomers. Kepler estimated the distance of the stars, which he supposed to be embedded in a hollow sphere with the earth as centre, as 420,000,000,000 miles. Newton estimated it as nine or ten millions of million miles. By the early nineteenth century measurement had greatly improved and still it remained impossible to detect any annual parallax—as the expected apparent motion of the near stars relative to the far ones was termed. The nearest stars were therefore well over twenty million million miles away and probably much further. In 1838 Bessel showed that the radius of the earth's orbit formed the base of an isosceles triangle whose vertex was the star 61 *Cygni* and contained an angle of rather less than a third of a second. This star was therefore about 60,000,000,000,000 miles from us. The vast majority of stars, however, both then and now, give no detectable annual shift and so must be much more distant than this.

What can we observe about stars too distant to give a parallax? Their light comes to us; so we can observe their spectrum and their brightness. Now the same body gives a slightly different spectrum at different temperatures, so if two stars give the same spectrum they must have about the same temperature, and also about the same size, for gravitational forces alter the spectrum somewhat. So if two stars have the same spectrum and one is 10,000 times brighter than the other, they may be supposed to give the same amount of light and the fainter to be $\sqrt{10{,}000}$, (i.e. 100) times more distant. If then the brighter is known to be so far off that light takes four years to come from it, then the fainter will be so distant that light will take 400 years to come from it.

Certain stars, called Cepheid variables, pulsate, i.e. grow alternately brighter and fainter, in a period of a few days, and the real brightness of these stars has been connected with their period of pulsation. So if an enormously distant and very faint star is seen to pulsate in the same period as a near and bright one, we know they are both giving out the same quantity of light and we know the ratio of their distances is the square root of the ratio of their brightness. This method has enabled us to measure the distances of enormously far-off stars.

Nebulæ and Galaxies

Among Sir William Herschel's many achievements was an attempt to estimate the size and shape of the body of stars visible to us, assuming that all stars are really of the same brightness and the difference in their apparent brightness is an effect of distance. Many more stars are visible in the plane of the circle of the Milky Way than at right angles to it and accordingly Herschel thus arrived at the notion of a roughly lens-shaped body of stars constituting the universe.

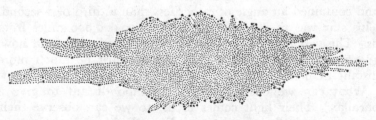

FIG. 47.—The shape of the universe according to Herschel (1785).

In the late eighteenth century note began to be taken of *nebulæ*, objects which appear as little hazy patches of light. Herschel studied these with his much improved telescopes and found that some of them could be resolved into clusters of stars, while others could not. As telescopes became more powerful it appeared that some of these nebulæ were really gigantic objects comprising countless stars, and all of them were at a greater distance than could be measured until recent years. When, in the twentieth century, measurements became possible it appeared that the nebulæ resolvable into stars were enormously further off than the stars in general and that they were, in fact, island universes like our own galaxy, each isolated by an unthinkable distance of space, empty of stars. Their form is commonly that of a flattened disc, like a biscuit, and there are considered to be about 100 million of them within the range of our greater telescopes (a sphere whose radius is about 500,000,000 light-years). Whether the universe continues so to infinity or whether there is any limit to it, we cannot tell. (See extract, p. 245.)

The spectra of these distant nebulæ show a 'red-shift' of the spectral lines, a decrease of frequency of the light given out by the glowing atoms, and science, for lack of a better explanation generally interprets this as implying that the whole universe is expanding, for there is no known cause for such a 'red-shift', other than velocity. We must not, however, place too much reliance on this, for who can tell what happens to light in its journey of 500,000,000 years from the furthest nebulæ? But if the nebulæ are racing away from each other they must once have been near each other and their motions indicate that they started some 1,800,000,000 years ago. What happened before this? We cannot tell, and the very belief that the nebulæ are receding depends only on our conjectural interpretation of their spectra.

The Evolution of Worlds

The earlier astronomers were impressed by the majestic permanence of the heavens and even Newton supposed God to have created the stars, sun, and planets, as they are now;

but towards the end of the eighteenth century, when men began to take a less literal view of the Scriptures, and to look for an explanation of the way the world might have reached its present state by the operation of known scientific laws, attempts were made towards a history of the heavenly bodies.

Immanuel Kant, the philosopher, was one of the first to consider the way in which the universe might have been formed from a uniform cloud of finely divided matter, and his theories are remarkably akin to those that are held to-day. Indeed he reached out too far for the science of his time and a more limited objective was proposed in 1796. Laplace, the greatest French astronomer of his time, put forward (fully realising its speculative character) a theory of the solar system which was discussed for many years (p. 240). He began with a vast mass of glowing nebulous matter in slow rotatory motion. As this mass cooled, it contracted; as it contracted the quantity of rotation remained the same and so the velocity of rotation increased. This continued until the velocity of the surface of the equator of the glowing mass became so great as to overcome the contrary force of gravity. A belt or ring of glowing matter thus separated from the equator of the mass—rather like one of the rings of Saturn—and after a time broke up into fragments which coalesced into a planet. This process repeated itself a number of times, so giving rise to all the planets. Saturn's rings and the circle of asteroids seemed to give physical evidence of its possibility. This, variously modified, was the general belief throughout the nineteenth century, but it has since become clear that there is not now, and presumably never can have been, sufficiently rapid rotation to account for the detachment of planets in this way: and all modern theories have to invoke the near approach to the sun of one or more stars, whose gravitational attraction drew out of it a mass of matter to be transformed into planets. Needless to say all such theories are hypotheses with but slender support; but where science cannot know, it is at liberty to conjecture, as long as conjectures are distinguished from facts.

Such a process as Laplace supposed had to be thought of as of gigantic duration, and as we have seen (p. 220) such men as

Hutton and the majority of geologists had proposed a vast age for the earth. But the earth could not be thought to be older than the sun, and after the Law of Conservation of Energy (pp. 322–24) was understood it was evident that the sun was giving out energy and this must have a source. If the Sun were simply a hot body in the process of cooling, or in some condition of chemical action (e.g. combustion), it could have supplied light and heat at its present rate for only a few centuries. Helmholtz in 1854 showed that the slow contraction of the sun could change potential gravitational energy into heat. This allowed a possible age of some twenty million years—but the geologists and biologists could not possibly account for the changes they studied in that time: to-day a thousand million years is not thought long enough for them. But until the twentieth century no better theory of the source of the sun's energy than that of Helmholtz could be suggested. In 1897 radio-activity was discovered and indirectly helped to give estimates of the age of the earth and sun. Some good evidence that the real age of the earth is two or three thousand million years has been supplied by the study of the proportions of uranium and an isotope of lead (into which it slowly changes) in the various rocks. Moreover, radioactivity showed us that the atoms of common matter were not inert lumps, but locked storehouses of incredible quantities of energy. To-day we suppose that the sun's energy is produced by the transmutation of hydrogen atoms into atoms of more complex character, a process which involves the annihilation of some of their mass, and a corresponding liberation of energy. The probable ages of the sun and earth as deduced (a) from considerations of energy-loss and replacement, (b) the evidence of radioactive change, (c) biological and (d) geological considerations are to-day in reasonable agreement.

Astronomers now conceive of the sun and of all stars as undergoing a process of irreversible change. They start with a vast tenuous and not very hot mass of gas, largely hydrogen, perhaps as large as might fill the entire orbit of Jupiter; gravitational attraction and consequent contraction causes a rise of temperature (as on Helmholtz's theory). When a certain temperature, reckoned in millions of degrees, is reached hydrogen atoms begin to build up into the atoms of heavier elements liberating energy. This process continues through thousands, or perhaps millions, of millions of years. The heat is dissipated by radiation from the surface of the star, but if at a slower rate than that at which it is formed within it, the temperature may rise and bring about explosive processes which are seen by us as the sudden appearance of a bright star where there was formerly a faint one. But whatever process occurs, it is irreversible. The star is losing energy and must become dark and cold: this has not yet occurred in the sun and stars we see: they must therefore have had a beginning in time about which science can tell us nothing. If the evolution of stars from formless matter began a million million years ago, we must ask what new causes, that were not operative a million million years before that, started those changes. We may think of Creation, we may be content to say we do not know, but we must not say that science has explained the beginning of things.

The Influence of Astronomy

The influence of Astronomical progress on practical affairs is not enormous. We are enabled by it to navigate and to tell the time, but its influence on our lives is negligible compared with that of chemistry or physics.

None the less, men have always regarded Astronomy as of intense interest, because it tells them what kind of place they live in, deduces its remote past, and dimly foreshadows its future. We have glanced at some of the changes in men's astronomical views and may sum them up by saying that, while the earth was always considered as minute, a mere point in comparison of the starry sphere, it was formerly thought of as a most important point, not only the seat of intelligence but also the geometrical centre of all. It is now seen by us as a geometrically insignificant speck among a stupendous multitude to which we can discover no limit: whether it is unique as a habitation of mind science cannot tell. Its history is no longer co-extensive with that of man. The weight of scientific evidence is against an infinitely extended past, but the past which we formerly reckoned as six thousand years cannot be shorter than 1800 million, and may be far longer.

What has man made of this? Generally, he has made the vulgar error of confusing size with excellence. Some have said to themselves: "How great must be the God that made this fabric!" Others: "How absurd that man, inhabitant of this speck, should believe the universe to be made for him!" Those accustomed to think of spiritual things have avoided these errors, realising that God is not in space, or time, but in infinity and eternity, and that miles and years need have no significance in His sight.

Examples Illustrating Modern Cosmologies

MILTON CONSIDERS THE NEW ASTRONOMY

And Raphael now, to Adam's doubt proposed,
Benevolent and facile, thus replied:
    "To ask or search, I blame thee not; for heaven
Is as the book of God before thee set,
Wherein to read his wondrous works, and learn
His seasons, hours, or days, or months, or years:
This to attain, whether heaven move or earth,
Imports not, if thou reckon right; the rest
From man or angel the great Architect
Did wisely to conceal, and not divulge
His secrets to be scanned by them who ought
Rather admire; or, if they list to try
Conjecture, he his fabric of the heavens
Hath left to their disputes, perhaps to move
His laughter at their quaint opinions wide
Hereafter; when they come to model heaven
And calculate the stars, how they will wield
The mighty frame! how build, unbuild, contrive
To save appearances; how gird the sphere
With centric and eccentric scribbled o'er,
Cycle and epicycle, orb in orb:
Already by thy reasoning this I guess,

Who art to lead thy offspring, and supposest
That bodies bright and greater should not serve
The less not bright, nor heaven such journeys run,
Earth sitting still, when she alone receives
The benefit. Consider first, that great
Or bright infers not excellence: the earth
Though, in comparison of heaven, so small
Nor glistering, may of solid good contain
More plenty than the sun that barren shines:
Whose virtue on itself works no effect,
But in the fruitful earth; there first received,
His beams, unactive else, their vigour find.
Yet not to earth are those bright luminaries
Officious; but to thee earth's habitant.
                    (*Paradise Lost*. Book VIII.)

[Note that Milton is considering the *purpose* of the parts of
the world: a question not considered by modern science, but
very important to the philosopher or man of religion.]

KANT'S ASTONISHING ANTICIPATION OF THE STRUCTURE OF THE UNIVERSE

I come now to that part of my theory which gives it its
greatest charm, by the sublime idea which it presents of the
plan of the creation. The train of thought which has led me
to it is short and natural; it consists of the following ideas. If
a system of fixed stars which are related in their position to a
common plane, as we have delineated the Milky Way to be,
be so far removed from us that the individual stars of which
it consists are no longer sensibly distinguishable even by the
telescope; if its distance has the same ratio to the distance of
the stars of the Milky Way as that of the latter has to the
distance of the sun; in short, if such a world of fixed stars is
beheld at such an immense distance from the eye of the spec-
tator situated outside of it, then this world will appear under
a small angle as a patch of space whose figure will be circular
if its plane is presented directly to the eye, and elliptical if it

is seen from the side or obliquely. The feebleness of its light, its figure, and the apparent size of its diameter will clearly distinguish such a phenomenon when it is presented, from all the stars that are seen single.

We do not need to look long for this phenomenon among the observations of the astronomers. . . . It is the 'nebulous' stars which we refer to, or rather a species of them, which M. de Maupertuis thus describes: 'They are,' he says, 'small luminous patches, only a little more brilliant than the dark background of the heavens; they are presented in all quarters; they present the figure of ellipses more or less open; and their light is much feebler than that of any other object we can perceive in the heavens.* . . .

It is far more natural and conceivable to regard them as being not such enormous single stars but systems of many stars, whose distance presents them in such a narrow space that the light which is individually imperceptible from each of them, reaches us, on account of their immense multitude, in a uniform pale glimmer. Their analogy with the stellar system in which we find ourselves, their shape, which is just what it ought to be according to our theory, the feebleness of their light which demands a presupposed infinite distance: all this is in perfect harmony with the view that these elliptical figures are just universes and, so to speak, Milky Ways, like those whose constitution we have just unfolded. And if conjectures, with which analogy and observation perfectly agree in supporting each other, have the same value as formal proofs, then the certainty of these systems must be regarded as established.

The attention of the observers of the heavens, has thus motives enough for occupying itself with this subject. The fixed stars, as we know, are all related to a common plane and thereby form a co-ordinated whole, which is a World of worlds. We see that at immense distances there are more of such star-systems, and that the creation in all the infinite extent of its vastness is everywhere systematic and related in all its members. . . .

The theory which we have expounded opens up to us a

* *Discours sur la Figure des Astres.* Paris, 1742.

view into the infinite field of creation, and furnishes an idea of the work of God which is in accordance with the infinity of the great Builder of the universe. If the grandeur of a planetary world in which the earth, as a grain of sand, is scarcely perceived, fills the understanding with wonder; with what astonishment are we transported when we behold the infinite multitude of worlds and systems which fill the extension of the Milky Way! But how is this astonishment increased, when we become aware of the fact that all these immense orders of star-worlds again iorm but one of a number whose termination we do not know, and which perhaps, like the former, is a system inconceivably vast—and yet again but one member in a new combination of numbers!* We see the first members of a progressive relationship of worlds and systems; and the first part of this infinite progression enables us already to recognise what must be conjectured of the whole. There is here no end but an abyss of a real immensity, in presence of which all the capability of human conception sinks exhausted, although it is supported by the aid of the science of number. The Wisdom, the Goodness, the Power which have been revealed is infinite; and in the very same proportion are they fruitful and active. The plan of their revelation must therefore, like themselves, be infinite and without bounds.

(*Universal Natural History and Theory of the Heavens*. Immanuel Kant. 1755. Translated W. Hastie. 1900.)

LAPLACE'S NEBULAR HYPOTHESIS

Thus to investigate the cause of the primitive motions of the planets, we have given the five following phenomena: 1st, The motion of planets in the same direction, and nearly in the same plane. 2nd, The motion of their satellites in the same direction, and nearly in the same plane with those of the planets. 3rd, The motion of rotation of these different bodies and of the Sun in the same direction as their motion of projection, and in planes but little different. 4th, The small excentricity of the orbits of the planets, and of their satellites.

* There is at present no evidence that these greater systems exist.

5th, The great excentricity of the orbits of comets, although their inclinations may have been left to chance . . . [after a discussion of Buffon's theory of the ejection of matter from the sun by the action of a comet, he continues]. Let us see if it is possible to arrive at their true cause.

Whatever be its nature, since it has produced or directed the motion of the planets and all their satellites, it must have embraced all these bodies, and considering the prodigious distance which separates them, it could only have been a fluid of immense extent. To have given them in the same sense a motion nearly circular about the sun, this fluid must have surrounded that luminary like an atmosphere. This view, therefore, of planetary motion, leads us to think that in consequence of excessive heat, the atmosphere of the Sun originally extended beyond the orbits of all the planets, and that it has gradually contracted itself to its present limits. In the primitive state in which we suppose the Sun, it resembled the nebulæ which the telescope shows us as composed of a more or less bright nucleus, surrounded by a nebulosity which, by condensing on the surface of the nucleus transforms it into a star. . . . [By considering these stars at an earlier and earlier stage of formation one arrives in the limit] at a nebulosity so diffuse that one could scarcely suspect its existence. . . . The atmosphere of the Sun cannot extend indefinitely: its limit is the point at which the centrifugal force due to its rotation balances its weight; but, as cooling condenses the atmosphere and condenses on the surface of the star the 'molecules' nearest to it, the velocity of rotation increases. . . . The centrifugal force due to this movement thus becoming greater, the point where the weight is equal to it is nearer to the centre. Supposing then, as is natural, that this atmosphere extended at any epoch to its limit, it was bound, as it cooled, to abandon the molecules situated at this limit and at successive limits produced by the increase of the rotation of the Sun. These abandoned molecules continued to circulate around that body, since their centrifugal force was balanced by their weight. . . . Now consider the zones of vapours successively abandoned. . . . The mutual friction of the molecules of each ring must have accelerated some and retarded others until they acquired the same angular motion. [They condense and approach the central part of the ring and], if the molecules continued to condense without breaking up, they would have finally formed a solid or liquid ring, but the regularity that this formation requires renders such a phenomena extremely rare. So the solar system offers only one example of it, the rings of Saturn. Almost always each ring of vapours must have broken up into several masses, which . . . have continued to circulate the same distance about the Sun. These masses must have taken a spheroidal form, with a movement of rotation directed in the sense of their revolution, since their lower* molecules have less real velocity than their higher ones: they have then as good as formed planets in the state of vapour. If one of these has been strong enough to reunite in turn all the others about its centre, the ring of vapours will have been thus transformed into a single spheroidal mass of vapours, circulating about the sun, with rotation directed in the sense of its revolution. [The four known asteroids he took as a case where these masses had not joined: satellites he supposed to be formed from planets by a process analogous to the formation of the planets themselves.] (*Exposition du Système du Monde*. Laplace. 6th Ed., 1835.) [The theory was put forward in a tentative way in 1796.]

HELMHOLTZ PROPOUNDS A THEORY OF THE SOURCE OF THE SUN'S HEAT

When the nebulous chaos† first separated itself from other fixed star masses it must not only have contained all kinds of matter which was to constitute the future planetary system, but also, in accordance with our new law,‡ the whole store of force which at a future time ought to unfold therein its wealth of actions. Indeed, in this respect an immense dower was bestowed in the shape of the general attraction of all the particles for each other. The force, which on the earth exerts

---

* Nearer the Sun.
† Assuming Laplace's theory of the history of the Sun and planets.
‡ Of the Conservation of Energy.

itself as gravity, acts in the heavenly spaces as gravitation. As terrestrial gravity when it draws a weight downwards performs work and generates *vis viva** so also the heavenly bodies do the same when they draw two portions of matter from distant regions of space towards each other.

The chemical forces must have been also present, ready to act; but as these forces can only come into operation by the most intimate contact of the different masses, condensation must have taken place before the play of chemical forces began.

Whether a still further supply of force in the shape of heat was present at the commencement we do not know. At all events, by aid of the law of the equivalence of heat and work, we find in the mechanical forces existing at the time to which we refer, such a rich source of heat and light, that there is no necessity whatever to take refuge in the idea of a store of these forces originally existing. When, through condensation of the masses, their particles came into collision and clung to each other, the *vis viva* of their motion would be thereby annihilated, and must reappear as heat. Already in old theories it has been calculated that cosmical masses must generate heat by their collision, but it was far from anybody's thought to make even a guess at the amount of heat to be generated in this way. At present we can give definite numerical values with certainty.

Let us make this addition to our assumption—that, at the commencement, the density of the nebulous matter was a vanishing quantity as compared with the present density of the sun and planets: we can then calculate how much work has been performed by the condensation; we can further calculate how much of this work still exists in the form of mechanical force, as attraction of the planets towards the sun, and as *vis viva* of their motion, and find by this how much of the force has been converted into heat.

The result of this calculation is, that only about the 454th part of the original mechanical force remains as such, and that the remainder, converted into heat, would be sufficient to raise a mass of water equal to the sun and planets taken together, not less than twenty-eight millions of degrees of the Centigrade

* Kinetic energy.

scale. . . . Neither is the Mosaic tradition very divergent, particularly when we remember that that which Moses names heaven, is different from the blue dome above us, and is synonymous with space, and that the unformed earth and the waters of the great deep, which were afterwards divided into waters above the firmament ·and waters below the firmament, resembled the chaotic components of the world:

'In the beginning God created the heaven and the earth.

'And the earth was without form and void; and darkness was upon the face of the deep. And the spirit of God moved upon the face of the waters.'

And just as in the nebulous sphere, just become luminous, and in the new red-hot liquid earth of our modern cosmogony light was not yet divided into sun and stars, nor time into day and night, as it was after the earth had cooled.

'And God divided the light from the darkness.

'And God called the light day, and the darkness He called night. And the evening and the morning were the first day.'

. . . Calculations show that, assuming the thermal capacity of the sun to be the same as that of water, the temperature might be raised to 28,000,000 of degrees, if this quantity of heat could ever have been present in the sun at one time. This cannot be assumed, for such an increase of temperature would offer the greatest hindrance to condensation. It is probable rather that a great part of this heat, which was produced by condensation, began to radiate into space before this condensation was complete. But the heat which the sun could have previously developed by its condensation, would have been sufficient to cover its present expenditure for not less than 22,000,000 of years of the past.

. . . Thus there are extinct suns. The fact that there are such lends new weight to the reasons which permit us to conclude that our sun also is a body which slowly gives out its store of heat, and thus will some time become extinct.

The term of 17,000,000 years* which I have given may perhaps become considerably prolonged by the gradual abate-

* Less than the 22,000,000 above mentioned, because the sun would previously have radiated more rapidly than it now does.

ment of radiation, by the new accretion of falling meteors, and by still greater condensation than that which I have assumed in that calculation. But we know of no natural process which could spare our sun the fate which has manifestly fallen upon other suns.

(Popular Lectures on Scientific Subjects. Lectures on the Interaction of Natural Forces. 1854. *On the Origin of the Planetary System.* 1871.
H. Helmholtz.)

THE LATEST VIEW

The homogeneity indicated by the reconnaissance, even as a rough approximation, is very significant. The uniform distribution extends out to the limit of our telescopes. There is no trace of a physical boundary, no evidence of a super-system of nebulæ isolated in a larger world. . . .

The picture suggested by the reconnaissance is a sphere, centred on the observer, about 1,000 million light-years in diameter throughout which are scattered about 100 million nebulæ. The nebulæ average about 85 million times as bright as the sun, their over-all diameters average between 15,000 and 20,000 light-years, and the average separation between neighbours is about 2 million light-years. A suitable model would be furnished by tennis balls, 50 feet apart, scattered through a sphere five miles in diameter.

(The Observational Approach to Cosmology. Edwin Hubble.
Clarendon Press, 1937. p. 18–19.)

CHAPTER FIFTEEN

The Age of Electricity

Static electricity

The scientists of the eighteenth century thought electricity to be interesting, but not useful or very important. Frictional electrical machines and Leyden jars could give shocks and sparks, but the quantity of electricity in them was extremely minute.

Frictional machines of the type that developed from von Guericke's (Pl. XV) gave the means of producing electricity continuously and at will. The real development of the subject, however, began only in 1729 when S. Gray discovered that the substances which are not electrified by friction, e.g. metals, could convey electricity for long distances, and du Fay, following this up, distinguished between conductors and insulators (Pl. XVIII, A). The idea of conductors implied the idea of electricity as something that flowed,—a *fluid*. He also showed that there were two kinds of electric fluid, like kinds repelling, and unlike attracting each other. Franklin put forward the theory in which these two kinds, positive and negative electricity, correspond to an excess or defect of one single fluid. Frictional machines gave only the minutest quantity of electricity, and a great advance was marked by the invention of the Leyden jar (Pls. XVI, D and XVII, D), which was a condenser capable of giving a considerable electrical discharge. The Leyden jar led to, indeed was discovered by means of, the electric shock. The powerful sparks it gave led Franklin to identify the electric spark and lightning (p. 154). During the last quarter of the century the quantitative study of electricity began, and the fact that the electrical attraction and repulsion followed the inverse square law was proved by Cavendish and Coulomb.

Electric batteries and their consequences

The physiological effect of electricity aroused great interest.

The investigation of the fact that a muscle-nerve preparation, as in a frog's leg, could be stimulated by electricity to contract, led to the discovery in the hands of Galvani and Volta, that the contact of two dissimilar metals was by itself enough to cause such a contraction; and Volta, following this up, discovered in 1800 the electric battery (Pl. XIX) which gave the means of studying currents of electricity. At once the science of electricity, as distinguished from its practical application, began to make rapid progress. Electrolysis was very soon discovered and led to the preparation of several new elements by Humphrey Davy, e.g. sodium and potassium (p 254). The heating effect of the current was at once obvious and in 1808 Davy demonstrated the first electric lighting in the form of a carbon arc (p. 258). It had long been noticed that magnetic effects were sometimes associated with lightning and in 1807 H. C. Oersted began a long research on the connection between the electric current and magnetic effects. The actual relationship was very different from what was expected, and it was not until 1819 that he discovered that a current flowing parallel to a compass-needle deflects it. Oersted thus discovered that the electric current had magnetic effects, and by 1825 electromagnets had been constructed. Electricity was not, however, of much use at this stage because, although very much better batteries than Volta's were constructed in the first forty years of the century, the electricity they made was extremely expensive. Yet there were two things the world wanted and that electricity could do as nothing else could, namely, silver-plating and long-distance communication; and these were its only important uses until the eighteen-seventies.

Electrolysis

As early as 1805 it was found that when a conducting object was connected to the negative pole of a battery and immersed in a solution of certain salts of metals, the positive pole being connected to a metal plate in the liquid, the metal deposited on the object. Electroplating thus became possible and was of commercial value from the eighteen-forties onward. The large scale use of electrolysis in chemical industry had to wait till the end of the century when electricity became cheap.

The telegraph

Perhaps the most striking property of electricity was its apparently instantaneous travel along an unlimited length of wire. The idea of signalling by means of the electric current—the electric telegraph—was tried out even before the electric battery was available, but it could not be of practical value, until batteries more reliable and steady than Volta's, and some means of rendering signals easily visible were discovered. In 1837 a magnetic telegraph was invented and applied in 1838 on the Great Western Railway. In 1847 the first public telegraph service was made available. In 1851 the first successful

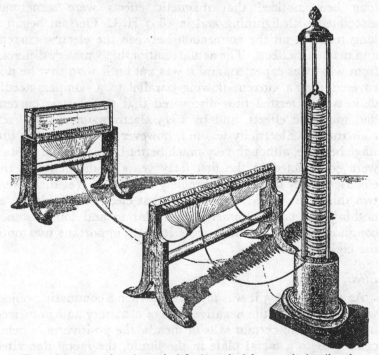

FIG. 48.—An early telegraph (1809) worked by a voltaic pile; the signals are indicated by electrolysis taking place at the terminals marked with the letters of the alphabet which it was desired to send.

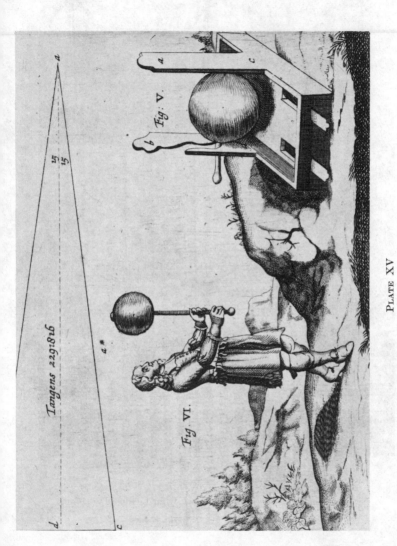

PLATE XV

The first electrical machine made by O. von Guericke about 1660.

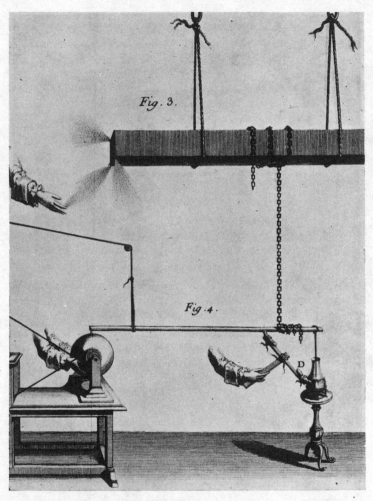

Fig. 3.

Fig. 4.

D

PLATE XVI
Brush discharge: birds killed by shock from Leyden Jar (1746).

submarine cable came into use. The telephone was invented in 1876, but although the first London exchange was opened in 1879 this means of communication did not become general until the twentieth century.

Electricity and Magnetism

That an electrical current could produce magnetic effects was discovered in 1819, and in 1831 Faraday discovered that the motion of a conductor in a magnetic field could produce an electric current. This has proved to be of gigantic practical and theoretical importance. The greatest practical result was the invention of dynamos, which generate electricity by moving coils of wire in a magnetic field (Pl. XX). Up to about 1870, these were comparatively inefficient converters of power into electricity, but as soon as Siemens made the discovery recorded on page 257, it became possible to use electricity for lighting and as a means of transmitting power. First came small private installations for lighting lighthouses or searchlights or work-shops. Next, about 1880, the first arrangements were made for public supply.

At first there were very few users of electricity, and so the generating stations were small and the electricity very expensive; but when electricity could be laid on to factories and houses, and was employed on a large scale for driving tramcars greater quantities had to be generated, and bigger and more efficient power-plant was used. As a result the cost fell from about half-a-crown a unit to the present industrial rate of a halfpenny or less, and this reduction of cost made further uses of electricity possible.

Electric lighting and power

Electric lighting created the first need for an electric supply. The electric arc had been known since 1807, and its extreme brilliance made it very suitable for street and factory lighting, though it was useless for domestic purposes. The carbon-filament lamp was invented in 1880–1; but it came into use but slowly, and in England was still something of a curiosity in the eighteen-nineties.

Meanwhile the electric motor had been considerably improved, and found a most important application in the electric tram-car. Road-transport was badly needed at the end of the nineteenth century. Towns had grown very large and were intensely overcrowded because the workman could not live at any distance from his work. Horse-trams were slow and small, so there was a welcome for electricity. The result was first of all to cause a great rehousing, for the cheapness of the tram-fares enabled workers to live in cottages five miles from their work, instead of in a single room—often a basement—in the middle of a city. The second result was to cause electricity to be generated on a large scale and therefore to become cheaper, and so to find a vast number of new users and uses.

Electric lighting gradually ousted gas-lighting from private houses in the years between 1900, when the former was universal, and 1920 when it was becoming rare. By this date electricity was cheap enough to be used for domestic heating and cooking, and its cost continued to fall until 1939.

Electricity is, of course, generated from coal, unless water-power happens to be available; and as only about a quarter of the energy of the coal can be turned into electrical energy, this must remain relatively expensive. None the less this is largely neutralised by its economical manner of use; for electricity can be very efficiently used in small independent power-units which can be switched on and off at will. Factories commonly contain thousands of machines. When they were run by steam, a complicated, wasteful and dangerous system of pulleys, belts, and shafting was needed to transmit the power from the steam-engine, which was the source of power, to the machines that used it. Much power was wasted in turning these shafts, especially when the machines were not in use, and in order to make this wastage a minimum, the factories were so designed as to require the least length of power-transmission. They were, therefore, commonly crowded, and of many storeys—and therefore tended to be unhealthy and dangerous. As their source of power was coal they had to be adjacent to railways. But in factories which are operated by electricity, each machine has its own motor and the only means of transmission is a wire.

Consequently, factories to-day are commonly built in the semi-rural outskirts of towns and consist of large one-storey buildings, abounding in space and light.

Electricity and Radiation

In the eighteenth century light was generally considered to be a stream of weightless particles, but even in the seventeenth century some had held the alternative view that, like sound, it was a wave-motion in a hypothetical 'luminiferous ether' which was supposed to fill all space. From 1800 this latter view gained acceptance and by 1840 was regarded as certain, for some phenomena such as polarisation and interference were readily explained on the wave-theory, but not easily otherwise. Two waves can cancel each other and give darkness, as happens with light, but two particles cannot. In 1849 the velocity of light was fairly accurately measured and it was noticed that the ratio of the electrical and magnetic units of electricity was equal to the velocity of light; it was also discovered that magnetism could affect polarised light. Faraday saw a connection between light, electricity and magnetism, and James Clerk Maxwell in 1864 worked this out mathematically and proved that light was a train of electromagnetic waves; which he considered to be very rapidly moving alternations of electrical and magnetic strain in the luminiferous ether. These alternations, though travelling at the same pace, might succeed each other at longer or shorter intervals, and Maxwell predicted that light was only one of a range of possible radiations of different frequencies. Infra-red and ultra-violet rays, akin to light but invisible, were already known, but what was required to clinch Clerk Maxwell's theory was to generate radiation by electrical means. This was done in 1887 by Hertz who generated by spark-discharges what we would now call very short radio-waves. This convinced the world that light was really electromagnetic. Other new types of radiation were soon discovered. X-rays were discovered in 1895; the gamma-rays of radium about 1900. Since then the whole range of frequencies from long-wave radio to the ultra short gamma-rays of radioactive substances has been produced and studied.

The gigantic importance of these cannot be exaggerated. The influence of broadcasting (from 1921, in the U.S.A.) is in itself a weighty factor in the history of the world: X-rays, if less universally known, have been the means of prolonging hundreds of thousands of lives.

Electricity as a constituent of matter

In the nineteenth century electricity seemed to be most mysterious. It moved like a material fluid, yet it seemed to have no mass and so to resemble immaterial force and energy. Yet despite the fact that science could form no picture of it, its effects were studied in very great detail. Normally the electric current travels through material conductors: but it can also be made to pass through empty space—or as near an approach to it as nineteenth-century vacuum-pumps could provide. Electricity crosses such a space in the form of rays, which were named cathode-rays, and in 1897 it was proved that these rays consisted of the minute negatively charged particles, which we now call electrons. Such work confirmed what had already been suspected, that, just as matter consisted of an enormous number of atoms, so electricity consisted of electrons, and was not a continuous fluid. Moreover electricity can be elicited from every chemical element, so all these elements clearly contained electrons. The atom was not, therefore, the simplest constituent of matter, and the electron was at least one constituent of every kind of atom. Here was, therefore, *a common basis for all kinds of matter* and the intuition of the ancient Ionian philosophers (p. 17) was a true one. A new science came into being—the study of the structures and internal mechanism of atoms. This, perhaps the most important development of the twentieth century, receives some mention in Chapter XIX.

Examples Concerning Electricity

MONSIEUR MONCONYS SEES THE FIRST
ELECTRICAL MACHINE

"On the twenty-second (of October, 1663) I went in the

morning to see M. Otho Gerike, Burgomaster and very learned in Pneumatics. . . .

He holds that the earth continually attracts all bodies to it, and to demonstrate this, he has a globe half-a-foot in diameter, made of what seemed to me a new mineral; it is yellowish and like highly polished cement*; when it was a little rubbed it attracted little leaves of certain vegetables, and the feathers of down; and what it is pleasing, it draws these feathers then lets them drop, then it picks them up and lets them drop again; and this continually and without end. . . ."

(*Les Voyages de Monsieur Monconys en Allemagne*. Paris. MDCXCV.)

EARLY EXPERIMENTS ON THE LEYDEN JAR

(There is some question as to who first discovered the Leyden jar: Musschenbroek, whose account follows, was long taken to be the discoverer, but von Kleist and possibly also Cuneus seem to have preceded him.)

I had suspended by two threads of blue silk an iron gun-barrel AB (Plate XVII) which received by communication the electricity of a glass globe, which was turned rapidly on its axis while being rubbed by application of the hands; at the end B there hung freely a brass wire, whose end was immersed in a round glass vessel D, partly filled with water, which I held in my right hand F, while with the other hand E, I tried to draw sparks from the electrified iron barrel: suddenly my right hand F was struck with such violence, that my whole body was shaken as if by a stroke of lightning . . . my arm and whole body were affected in a terrible manner which I cannot express: in a word I thought I was done for.

(THE BRUSH DISCHARGE AS DESCRIBED BY THE ABBÉ NOLLET)

M. l'Abbé Nollet thought that he would be more readily successful if he employed a much stronger electricity: so instead

---

* Actually of sulphur. It is illustrated in Plate XV.

of electrifying, as was usual, a rod of iron or a gun-barrel, he electrified a square bar of iron weighing 60–80 lbs. . . . From the four angles of this bar there were seen to issue four burning jets (*gerbes*) more than five inches long, and the noise that they made could be heard in the next room of which the door remained open; at a distance of more than fifteen inches from the bar, one felt a very considerable and very sensible wind. A finger held four inches from the bar, became luminous at the tip; there issued from it a little brush (*aigrette*). (Plate XV).

THE MORTAL EFFECT OF THE ELECTRIC DISCHARGE

The event justified his fear; two small birds attached to either end of a metal rule, were so held that one touched the glass vessel while the other was brought up to the bar to draw from it the spark: scarcely had it reached two inches from the bar when there issued from it a flash of fire which struck it with such violence that it gave hardly any sign of life, and at a second stroke, it remained quite dead. . . . (Plate XVI).

(*Histoire de l'Academie Royale des Science* avec les Mémoires de Mathematique et de Physique pour le même année (1746). The first extract is from p. 2 of the *Memoires*: the latter two from pp. 9–10 of the *Histoire*).

HUMPHRY DAVY DISCOVERS POTASSIUM BY ELECTROLYSIS

A small piece of pure potash, which had been exposed for a few seconds to the atmosphere, so as to give conducting power to the surface, was placed upon an insulated disc of platina, connected with the negative side of the battery of the power* of 250 of 6 and 4, in a state of intense activity; and a platina wire, communicating with the positive side, was brought in contact with the upper surface of the alkali. The whole apparatus was in the open atmosphere.

Under these circumstances a vivid action was soon observed

---

* Of 250 plates of 6 inches and 4 inches square. No units for expressing E.M.F. or current were yet devised.

to take place. The potash began to fuse at both its points of electrization. There was a violent effervescence at the upper surface; at the lower, or negative surface, there was no liberation of elastic fluid;* but small globules having a high metallic lustre, and being precisely similar in visible characters to quicksilver, appeared, some of which burnt with explosion and bright flame, as soon as they were formed, and others remained, and were merely tarnished, and finally covered by a white film which formed on their surfaces.

(*Bakerian Lecture to the Royal Society.* Humphry Davy. 1807.)

DAVY DESCRIBES THE ELECTRIC ARC

The most powerful combination† that exists in which number of alternations is combined with extent of surface,‡ is that constructed by the subscription of a few zealous cultivators and the patrons of science in the laboratory of the Royal Institution. It consists of two hundred instruments, connected together in regular order, each composed of ten double plates arranged in cells of porcelain and containing in each plate thirty-two square inches; so that the whole number of double plates is 2,000 and the whole surface 128,000 square inches. This battery, when the cells were filled with 60 parts of water mixed with one part of nitric acid, and one part of sulphuric acid, afforded a series of brilliant and impressive effects. When pieces of charcoal about an inch long and one sixth of an inch in diameter, were brough near each other (within the thirtieth or fortieth part of an inch), a bright spark was produced, and more than half the volume of the charcoal became ignited to whiteness, and by withdrawing the points from each other a constant discharge took place through the heated air; in a space equal at least to four inches, producing a most brilliant ascending arch of light, broad and conical in form in the middle. When any substance was introduced into this arch, it instantly became ignited; platina melted as readily in it as wax in the

* i.e. gas.
† Electric battery.
‡ Alternations=cells in series. Such a battery would have an E.M.F. of c. 200 volts and give a very large current when first connected.

flame of a common candle; quartz, the sapphire, magnesia, lime all entered into fusion; fragments of diamond, and points of charcoal, and plumbago rapidly disappeared, and seemed to evaporate in it even when the connection was made in a receiver exhausted by the air-pump; but there was no evidence of their having previously undergone fusion.

(*Elements of Chemical Philosophy*. Humphry Davy. 1812.)

LIGHT AN ELECTRO-MAGNETIC WAVE

[Faraday in 1845 showed that a magnetic field was capable of rotating a ray of polarised light.]

Thus is established, I think for the first time, a true direct relation and dependence between light and the magnetic and electric forces; and thus a great addition made to the facts and considerations, which tend to prove that all natural forces are tied together and have one common origin . . . the great power manifested by particular phænomena in particular forms is here further identified and recognised by the direct relation of its form of light to its forms of electricity and magnetism.

(*Philosophical Magazine*. Nov. 1845.)

[James Clerk Maxwell in 1862 worked out the relationship of electricity, magnetism and light. He supposes a medium, the luminiferous ether, which is in a state of rapid alternations of magnetic and electrical strain. From this theory he calculates the speed at which these alternations or waves ought to propagate themselves.]

By the electromagnetic experiments of MM. Weber and Kohlrausch

$$v = 310,740,000 \text{ metres per second}$$

is the number of electrostatic units in one electromagnetic unit of electricity, and this, according to our result, should be equal to the velocity of light in air or vacuum.

The velocity of light in air* by M. Fizeau's experiments, is

$$V = 314,858,000;$$

* In m. per second.

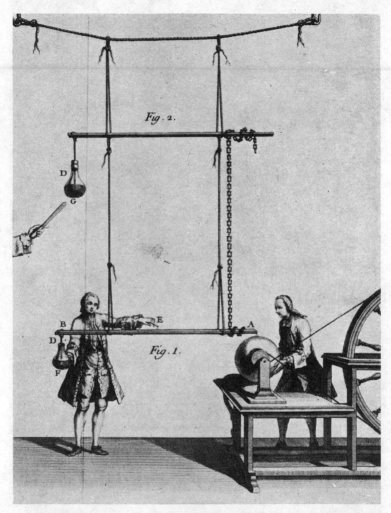

Fig. 2.

Fig. 1.

PLATE XVII

Charging a Leyden Jar by means of a frictional machine (1764).

PLATE XVIIIa

Demonstration that the conduction of electricity is almost instantaneous (1746).

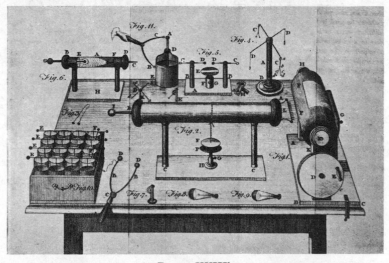

PLATE XVIIIb

Various pieces of electrical apparatus used at the end of the eighteenth century.

according to the more accurate experiments of M. Foucault

$$V = 298,000,000$$

The velocity of light in the space surrounding the earth, deduced from the coefficient of aberration and the received value of the earth's orbit, is

$$V = 308,000,000.$$

Hence the velocity of light deduced from experiment agrees sufficiently well with the value of v deduced from the only set of experiments we as yet possess. The value of v was determined by measuring the electromotive force with which a condenser of known capacity was charged, and then discharging the condenser through a galvanometer, so as to measure the quantity of electricity in it in electromagnetic measure. The only use made of light in the experiment was to see the instruments. The value of V found by M. Foucault was obtained by determining the angle through which a revolving mirror turned, while the light reflected from it went and returned along a measured course. No use whatever was made of electricity or magnetism.

The agreement of the results seems to show that light and magnetism are affections of the same substance and that light is an electromagnetic disturbance propagated through the field according to electromagnetic laws.

(*A Dynamical Theory of the Electromagnetic Field.* J. Clerk Maxwell. *Philosophical Transactions,* 1864.)

THE IDEA OF ELECTRIC POWER AND SUPPLY

[Sir W. Siemens writes to his brother William, on December 4th, 1866.]

I have had a new idea, which, in all probability will succeed and will give important results.

As you well know Wilde has taken a patent in England consisting in the combination of a magnet-inductor* of my construction with a second one which has a large electromagnet instead of the steel magnet. The magnet inductor as constructed in our alphabetical telegraph instruments, magnetises

* i.e. a dynamo with permanent steel magnets.

the electro-magnet to a higher degree than can be obtained by steel magnets. The second inductor will therefore give much more powerful currents than if it had steel magnets. The action ought to be colossal, as is stated in *Dingler's Journal*.

But now, clearly, the magnet-inductor with steel magnets may be entirely dispensed with. If we take an electro-magnetic machine, which is so constructed that the stationary magnet is an electromagnet with a constant polar direction, while the current of the movable magnet is changed; and if we insert a small battery,* which will thus work the apparatus, and now turn the machine in the contrary direction the current must increase. The battery may (then) be excluded and removed . . .

We may thereby with the sole aid of wire-coilings and soft iron transform power into current.

. . . Magneto-electricity will by this means become cheap and electric lighting, galvanometallurgy,† and even small electromagnetic machines‡ receiving their power from larger ones will become possible and useful.

(*Life of Sir W. Siemens*. W. Pole. 1888. p. 233.)

SIR J. J. THOMSON FINDS A COMMON CONSTITUENT OF ALL MATTER

[J. J. Thomson's study of cathode-rays convinced him that they were negatively-charged particles whose mass was much less than that of an atom and that these were identical whatever their source. In 1897 he writes:]

Thus on this view we have in the cathode-rays, matter in a new state, a state in which the sub-division of matter is carried very much further than in the ordinary gaseous state: a state in which all matter—that is, matter derived from different sources, such as hydrogen, oxygen, etc.—is of one and the same kind, this matter being the substance from which the chemical elements are built up.

(*Cathode Rays*. J. J. Thomson. *Philosophical Magazine*. XLIV. p. 312.)

---

* This proved to be unnecessary.
† i.e. electro-plating, large-scale electrolysis, etc.
‡ i.e. Electric motors.

[In 1899:]

I regard the atom as containing a large number of smaller bodies which I will call corpuscles; these corpuscles are equal to each other; the mass of a corpuscle is the mass of a negative ion* in a gas at low pressure, i.e. about 3×10^{-28} gm. In the normal atom this assemblage of corpuscles forms a system which is electrically neutral. Though the individual corpuscles behave like negative ions, yet when they are assembled in a neutral atom the negative effect is balanced by something which causes, the space through which the corpuscles are spread, to act as if it had a charge of positive electricity equal in amount to the sum of the negative charges on the corpuscles.

(*The Masses of Ions in Gases at Low Pressures.* J. J. Thomson. *Philosophical Magazine.* XLVIII. p. 565.)

---

* The distinction between what we now call an ion, i.e. an electrified atom or molecule, and an electron, a particle of electricity, was not yet clear. Here Thomson intends what we now call an electron.

CHAPTER SIXTEEN

The Acceleration of Transport

Transport in the eighteenth century

Transport is required for goods and passengers. It is desirable that it should be reliable, safe, punctual, cheap and speedy. For heavy goods-traffic, the chief desideratum is low cost; for passenger-traffic, high speed.

In the early eighteenth century none of the above qualities were possessed by any of the means of transport. On land goods were carried in panniers or in heavy stage-waggons, while passengers travelled on horseback or in coaches, which, at this period, owing to the state of the roads, travelled on the average not much faster than a walking-pace. By sea, the sailing-ships were at the mercy of the winds; a journey to Lisbon might take a fortnight or a couple of months—no one could tell. The slowness of travel made it expensive, for other things being equal, the services of men and materials were required for a far longer time than is the case to-day. Travel was unsafe owing to the depredations of highwaymen and pirates, coach-accidents were not infrequent, and sailing-ships were far more subject to the perils of storm than is the modern liner.

These defects in the transport of the time were realised, and became very conspicuous when the growing industries began to produce and distribute large quantities of goods. The means of improvement was found in *diminishing the amount of work wasted in friction.*

In the latter half of the eighteenth century men found out how to make good, hard, even roads; thus allowing coaches to move faster, and stage-waggons, which never exceeded a walking-pace, to use fewer horses to draw a given weight of goods. At the same time canals were constructed over a great part of England, so that great weights of heavy goods could be drawn by a single horse towing a barge; at about the same period the first iron railways were constructed, mostly in collieries. These were only horse-railways, but they demonstrated how much greater was the weight that a given force could move on rails than on even the best road. So at the close of the eighteenth century, passenger-traffic might in favourable circumstances attain eight miles an hour, and goods-traffic, though it never exceeded three or four miles an hour, could be carried with much less expenditure of labour than previously and therefore at much lower cost.

The Locomotive Engine

The development of the steam-engine is dealt with in Chapter X. All the engines before James Watt's were enormously heavy and bulky, and could not have been used to propel a locomotive. Watt's improvements greatly lightened the steam-engine, but the one which made the locomotive possible was the use of the excess of steam-pressure over atmospheric pressure to drive the piston, instead of relying on the excess of atmospheric pressure over that of the partial vacuum in the condenser. The steam-engine, then, could dispense with the heavy and bulky condensers (though at the cost of losing some of its efficiency) and thus was made light enough to be mounted on a vehicle. Between 1801 and 1825 quite a number of locomotives were constructed and used for haulage on private colliery railroads. They were designed to pull heavy trains of trucks at a slow pace, and the important step of adapting the locomotive to speedy passenger transport was made in 1830 by George Stephenson, who had had considerable experience in building the earlier haulage-locomotives. His prize engine, the "Rocket", built for the Liverpool and Manchester Railway, averaged fourteen miles an hour, but achieved thirty-six: probably twice as fast as any wheeled traffic had ever travelled before. The essential features of the locomotive engine were:

(1) Abandonment of the condenser;

(2) The carrying of the exhaust-steam into the chimney so producing a strong draught in order to cause the fuel to burn rapidly.

The first condition, and occasionally the second, were fulfilled in engines earlier than the Rocket, the chief merit of which, apart from general good design and workmanship, was:

(3) The use of a boiler fitted with tubes carrying the fire-gases through it, so enabling great quantities of steam to be rapidly produced in a small boiler.

Progress in Railways

The Liverpool and Manchester Railway aroused the enthusiasm of the civilised world, whose capital resources were soon strained to the utmost in covering England, Europe, and America with a network of railways. In this country through-travel to most important places was possible by 1850–60. In 1872 the great Railway Companies were formed by extensive amalgamation of minor lines and at the same period third-class travel at a penny a mile and on all trains began to come into operation. Workmen's tickets were first issued in 1883, and this made it possible for the better-paid workers to live in suburbs instead of in the heart of the great towns. London consequently extended rapidly, and a part of her population was freed from the squalor of overcrowded tenements. From about 1870 cheap steel was available and steel rails replaced those of wrought-iron. This, and the Westinghouse brake (1868) enabled heavier and faster trains to be run. Sleeping-cars and dining-cars appeared in the 'seventies, corridor-trains and lavatory accommodation about 1890. Intense competition between railway companies steadily raised the standard of facilities, but no other form of transport competed with the railway until after the war of 1914–1918, when road transport became cheap and comfortable. (Plate XXV.)

Steamships

Locomotives were invented for goods-haulage and only later came into use for passenger-transport. The contrary was the case with steamships.

It was not difficult to adapt a steam-engine to propel a ship, because the weight and bulk of the engine was not an important matter. Some difficulty was found in modifying ship-design to withstand the stresses imposed by a steam-engine, but the chief obstacle to the widespread adoption of steamships was the cost of coal. Land-transport was expensive, so the cost of fuel was not an important factor, but sea-transport was very much less expensive, and consequently sail competed with steam until near the end of the nineteenth century. Thus, until about 1870, fast passenger-boats used steam, but cargo-boats used sail. But in the 'sixties and 'seventies great improvements were made in marine engines. The old low-pressure engine, driving a paddle-wheel direct, was replaced by a high-pressure compound engine which was far more economical and was suitable for driving a propeller. Turbines came into use in large high-speed vessels from about 1900. We may sum up the change by saying that in 1850 only about 5% of the world's tonnage was mechanically propelled: in 1883 this had risen to 50% and in 1923 to 95%.

Electric Transport. (See Chapter XV.)

Electricity cannot be usefully employed at present for sea-transport, though the invention of small, light and capacious storage-batteries might at any time render this a possibility. Its importance for land-transport has been growing since about 1890. It is used for road-transport (tram-cars and trolley-buses), for underground railways, in which for obvious reasons it is likely to retain its present monopoly, and for electric railway-trains, which have great advantages for suburban work where much stopping and starting is required. The power used is of low cost because it is very efficiently generated, but the capital cost of overhead wires and track is considerable and the route, once laid down, cannot be varied at will. The electrification of railways is steadily advancing, but in this country electric road-transport reached its peak in 1929 and has since receded in favour of motor-transport.

The Internal Combustion Engine and Road Transport

From about 1825–1850 steam-carriages were used on the roads. They appear to have been efficient, though very heavy and smoky. They were opposed both by the railways and by those who used horse-carriages, and the high turnpike duties drove them off the roads. But even on the Continent, where such restrictions did not operate, these steam-carriages were not much favoured. They weighed three tons or more and had all the inconveniences of furnaces and boilers. The roads were, moreover, not good enough to allow solid-tyred vehicles to proceed at more than about ten miles an hour.

Internal combustion engines, designed to run on gas, were invented quite early in the nineteenth century, but not till 1876 was the four-stroke gas-engine of the modern type put on the market. This was easily adapted to burn an inflammable vapour, and there is nothing in the petrol-engine that could not have been constructed at a much earlier date—say 1855; but until a suitable volatile fuel was available at low cost there was no object in designing such an engine.

When mineral oil began to be distilled, in order to make paraffin (kerosene) for lamps, petrol (then called benzoline or benzine*) became available as a by-product. It thus became practicable to burn petrol-vapour in a modified gas-engine, and in 1887 Daimler invented his petrol-engine, which differed from every previous engine by reason of its lightness. It weighed 88 lbs. per horse power, while steam-engines weighed 300 lbs. per horse power or even more. Later designs were made lighter still, and to-day aeroplane engines may weigh a pound or less for each horse power.

The petrol-engine, then, was the instrumental cause of the two great developments in transport which characterise the twentieth century, namely, motor-transport and air-transport. It took a very long time to find out how to design a reliable, high-speed vehicle and to persuade the local authorities to make roads it could use; consequently it was not until about 1909 that motoring became a useful means of transport and not merely a way of making pleasure-trips. In this year, too, the motor-bus began to compete seriously with the horse-bus and the electric tram. After 1920 the motor-coach and private car became serious competitors of the railways.

* Distinguish from *benzol* or *benzene*, a coal-tar product.

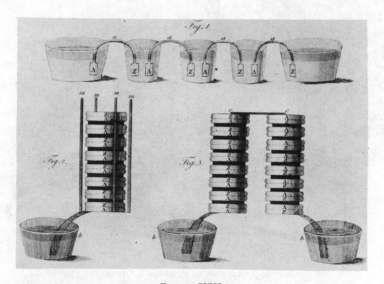

PLATE XIX

Volta's batteries. Z are zinc plates and A are copper plates, separated
by acid either as in glass vessels (above) or soaked up in porous
fabric (below).

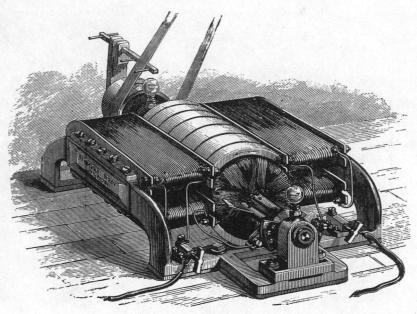

PLATE XX
An early dynamo, as used for lighting.

PLATE XXI
Lunardi's balloon (1784).

PLATE XXII

The apparatus used to generate the hydrogen required for Lunardi's
balloon from diluted sulphuric acid and iron nails.

Flying

Balloon-trips had been a marvel and excitement since 1783, but, except for the making of scientific observations, they were of little use. (Pls. XXI, XXII.) The idea of a dirigible balloon was a favourite one in the nineteenth century, and as soon as a light and powerful engine was available it became possible to make one. In 1900 Graf von Zeppelin operated his first airship successfully. In the early days of flying the airship was more trustworthy than the aeroplane, but in the last twenty years the position has been reversed.

Model aeroplanes were flown as early as 1848, and in 1894 Sir Hiram Maxim's large steam-driven plane managed to rise a few inches from the ground. The first aeroplane, as we know it, was made by Wilbur and Orville Wright in 1903, when they added a petrol-motor and air-screw to the glider which they had previously learnt to fly (Pl. XXIIIa). The years 1914–18 saw the aeroplane made a reliable machine that anyone might learn to operate: the years 1920–39 saw a network of air-services established all over the globe. To-day America is not half a day away from us. Finally, the years 1939–45 have demonstrated the fearful power of the aeroplane as a weapon of war.

Reaction-Motors

The latest development is the application of the principle of the rocket to the aeroplane. The propeller is dependent upon air against which to thrust, but the rocket and the jet-plane depend only upon the ejection of gases and would move as rapidly in a vacuum. The invention of the jet-plane therefore opens the way to travel in sealed planes at very great heights where weather is non-existent and air resistance very slight. It is therefore likely to provide a means of travelling at more than a thousand miles an hour.

Rocket propulsion has already afforded the means of propelling projectiles at speeds approaching 3,000 miles per hour with reasonable accuracy as was shown by the rocket-bomb attacks on London in 1944–5. There is little doubt that the range and accuracy of these are being greatly increased and that when loaded with an atomic bomb they constitute the most menacing form of war-material at present known.

Influence of Mechanical Transport

A system of reasonably rapid and cheap goods-transport is something without which our industrial civilisation could not have developed. Almost any kind of large-scale production requires the means of bringing coal and raw material to the works and of taking the finished products away. Factory-production was in fact the original cause of the search for new means of transport, and led directly to the building of the railways. Furthermore the great lowering of the cost of transport which science has occasioned, means that all but the heaviest and cheapest goods can be taken hundred or thousands of miles from the factories where they are made and sold at prices which compete favourably with those charged by local manufacturers. Thus the falling cost of transport has led to growing world-trade: and later to the concentrating of manufactures in very large and efficient factories, so saving in cost of production more than smaller and less efficient local factories can save by dispensing with the need for long-distance transport.

Mechanised transport, being cheap and speedy, has begun to equalise the world's food-resources. Before 1800 a failure of crops extending over a kingdom meant death by starvation for thousands. Nowadays, in time of peace, it would mean only a temporary slight rise of prices; for food would be imported, perhaps from another continent. Transport has therefore greatly depressed agriculture in countries where food cannot be cheaply grown, and has turned Canada, much of the U.S.A., and Australia into food-producers for the world, and, in turn, has made their people highly dependent on the prosperity of the foreign countries that are their customers.

Foreign travel has increased beyond measure. In the eighteenth century few but the wealthy and leisured went abroad; to-day, the man or woman of thirty who has not crossed the Channel is exceptional. The habit of yearly holidays, unknown before the eighteen-forties, and far from general until the 'eighties, arises from rail-traffic. It would not be difficult to indicate a great number of such minor changes in our lives wrought by transport, but all such look very small indeed beside the gigantic fact that *transport made our industrial civilisation possible.*

The Consequences of Road-transport

Road transport before 1880 was of very little consequence except in big cities and in places which the railway did not reach. In these cities there were commonly horse-omnibuses, in the country the carrier's cart was often the only public conveyance. The first bicycles came in the 'sixties, but cycling was not a practical or popular way of travelling until about 1896, when it became almost universal.

By this time the electric trams were installed in many cities, but these of course did not ply much beyond the city boundaries.

The early motor-buses were likewise town-vehicles and it was only from 1919 onwards that the towns and villages of England became linked by motor-bus and coach services and that the roads became filled and congested with private motor-cars.

The effect of this road-transport, broadly speaking, has been to get people out of the towns and greatly to reduce overcrowding. The cheap rail-fares of the 'eighties took many of the more prosperous workers out of the slum-tenements to an inner ring of suburbs: the electric trams of nineteen hundred took them to the outer suburbs and left in the central slums only those who were forced to dwell there by extreme poverty or by the nature of their work. The cheap and plentiful road-transport of the years since 1920 has moved the better-off workers into the villages scattered near the towns, and has enabled most of the slums to be pulled down and replaced by housing-estates, from which even the poorest can afford to travel to work. This has been an enormous boon from the point of view of health and self-respect, but it is to be remembered that to-day there are too many workers who waste an exhausting and unpaid hour or hour and a half in daily travel. The remedy for this is the replanning of road and rail-routes.

Transport and War

The size of armies was formerly limited by the difficulties of feeding them, and until the age of railways they rarely exceeded 200,000 men. The use of railways for military transport began about 1860 and although invasion by railway was not possible, the railways permitted armies of millions to be supplied. In the war of 1914–18 motor-transport was not very far advanced and was practically limited to the roads, but the characteristic of the recent war was the ability to move armies and their supplies over any but the most mountainous country at the rate of fifty miles or more a day. We have begun to realise the possibility of moving considerable armies by air: it would seem that in a few years' time they will be able to be permanently supplied by air and natural barriers and frontiers will cease to have any significance. The world will have to take precautions against this new threat of large-scale invasion at ten minutes' notice, but it is unlikely to do so until it is too late.

Examples Illustrating the History of Transport

COACHES IN 1692

There is an admirable commodiousness both for men and women of the better rank to travel from London, the like of which has not been known in the world; and that is, by stage coaches, wherein one may be transferred to any place, sheltered from foul weather, with a velocity and speed equal to the fastest posts* in foreign countries for the stage-coaches called 'Flying coaches' make forty or fifty miles a day.

(This would refer to fine-weather travel; in winter the roads sometimes became impassable.)

"FLYING-COACH" ADVERTISEMENT, 1775

HEREFORD MACHINE. In a day and a half, twice a week, continues flying from the Swan in Hereford, Monday and Thursday to London.

(Hereford is about 150 miles from London.)

* i.e. relays of riding-horses.

RICHARD TREVITHICK TRIES HIS FIRST LOCOMOTIVE
ON THE ROAD

I knew Captain Dick Trevithick very well; he and I were born in the same year. I was a cooper by trade and when Captain Dick was making his first steam-carriage I used to go every day into John Tyack's blacksmith's shop at the Weith, close-by here, where they put her together.

The castings were made down at Hayle, in Mr. Harvey's foundry. There was a deal of trouble in getting all the things to fit together. Most of the smith's work was made in Tyack's shop.

In the year 1801, upon Christmas Eve, coming on evening, Captain Dick got up steam, out in the high-road, just outside the shop at the Weith. When we see'd that Captain Dick was agoing to turn on steam, we jumped up as many as could; may be seven or eight of us. 'Twas a stiffish hill going from the

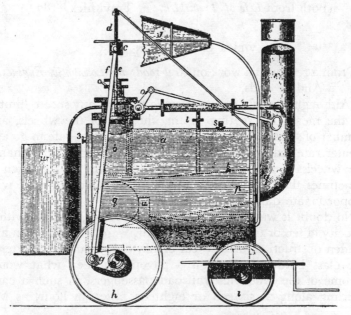

FIG. 49.—Trevithick's locomotive. Note vertical cylinder and bellows to provide draught.

Weith up to Camborne Beacon,* but she went off like a little bird. When she had gone about a quarter of a mile, there was a roughish piece of road covered with loose stones; she didn't go quite so fast, and as it was a flood of rain, and we were very squeezed together, I jumped off. She was going faster than I could walk, and went on up the hill about a quarter or half a mile farther, when they turned her and came back again to the shop. Captain Dick tried her again the next day; I was not there, but heard say that some of the castings broke.

(Old Stephen Williams' account, given in 1858 when he was 87.)

"The next day it went down to Crane, a short mile, that Captain Andrew Vivian's family, who lived there, might see it. Old Mrs. Paul cried out 'Good gracious, Mr. Vivian! What will be done next? I can't compare 'un to anything but a walking puffing devil'."

(Captain Joseph Vivian's account.)

(Both from *Life of Trevithick*. F. Trevithick. 1872.)

VIEWS ON SPEED IN 1825

(From a review of a work on *Rail-roads and Locomotive Engines*.)
. . . And he adds,
'Although it would be practicable to go at any speed, limited by the means of creating steam, the size of the wheels, and number of strokes in the engine, it would not be safe to go at a greater rate than nine or ten miles an hour. If by any chance the wheels of the engine should get off the rails, which is sometimes the case, a greater speed would be attended with proportionate danger.'

No doubt it would; for if ponderous bodies, moving with a velocity of ten or twelve miles an hour, were to impinge on any sudden obstruction, or a wheel break, they would be shattered like glass bottles dashed on a pavement: then what would become of the Woolwich railroad passengers, in such a case, whirling along at sixteen or eighteen miles an hour, as Mr.

* The road has since been diverted, but it is known that the incline was such that horse-vehicles had to proceed at a walking pace.

Telford says, 'with greater safety' than the ordinary coaches? We trust, however, that Parliament will, in all the railroads it may sanction, limit the speed to eight or nine miles an hour.

(*Quarterly Review*. Vol. XXXI, 1825, pp. 368–9.)

TRANSPORT BEGINS TO TAKE ITS TOLL

(Opening of the Manchester and Liverpool Railway in 1830.)

At Parkside, about seventeen miles from Liverpool the engines stopped to take in water. Here a deplorable accident occurred to one of the illustrious visitors, which threw a deep shadow over the subsequent proceedings of the day. The "Northumbrian" engine with the carriage containing the Duke of Wellington, was drawn up on one line, in order that the whole of the trains on the other line might pass in review before him and his party. Mr. Huskisson had alighted from the carriage and was standing on the opposite road along which the "Rocket" was observed rapidly coming up. At this moment the Duke of Wellington, between whom and Mr. Huskisson some coolness had existed, made a sign of recognition and held out his hand. A hurried but friendly grasp was given; and before it was loosened there was a general cry from the bystanders of "Get in, get in!" Flurried and confused, Mr. Huskisson endeavoured to get round the open door of the carriage which projected over the opposite rail; but in so doing he was struck down by the "Rocket", and falling with his leg doubled across the rail, the limb was instantly crushed. His first words, on being raised, were "I have met my death", which unhappily proved true, for he expired that same evening in the parsonage of Eccles. It was cited at that time as a remarkable fact that the "Northumbrian" engine, driven by George Stephenson himself, conveyed the wounded body of the unfortunate gentleman a distance of about fifteen miles in twenty-five minutes, or at the rate of thirty-six miles an hour. This incredible speed burst upon the world with the effect of a new and unlooked for phenomenon.

(*Lives of the Engineers*. Vol. III. Samuel Smiles. 1862.)

ERASMUS DARWIN (1791) ANTICIPATES STEAMSHIPS, LOCOMOTIVES AND AIR-WARFARE

Soon shall thy arm, UNCONQUERED STEAM! afar
Drag the slow barge, or drive the rapid car;
Or on wide-waving wings expanded bear
The flying-chariot through the fields of air.
Fair crews triumphant, leaning from above,
Shall wave their fluttering kerchiefs as they move;
Or warrior-bands alarm the gaping crowd,
And armies shrink beneath the shadowy cloud.

(*The Botanic Garden*. Erasmus Darwin. 1791.)

THE BOMBING OF CITIES FORESEEN IN 1912

The point to be considered, in this connection, is this: such an aerial attack is no longer a vague possibility. It was only the other day, while discussing the destructive capabilities of modern type aeroplanes, that a famous constructor showed how—if a large fleet of machines was marshalled together—it would be possible for an enemy to drop a couple of hundred tons of explosive matter upon London, suddenly appearing from across the Channel by air, and flying as quickly back again.

What such an aerial attack as this would mean has been pictured by Lord Montagu of Beaulieu. Suppose London was thus assailed from the air at the beginning of a war, he says: What would the result be? Imagine the Stock Exchange, the chief banks,* the great railway stations and our means of communication destroyed. "Such a blow at the very heart of the Empire," declares Lord Montagu, "would be like paralysing the nerves of a strong man, with a soporific, before he had to fight for his life: the muscular force would remain, but the brains would be powerless to direct." . . . In conclusion it may be taken that the offensive possibilities of the aeroplane grow from day to day. Machines are built to fly faster and to carry heavier weights. In future so far as the question of this destructive

* Note, as typical of the Edwardian era, the vast importance attached to the Stock Exchange and banks.

PLATE XXIIIa
The Wrights' aeroplane in flight.

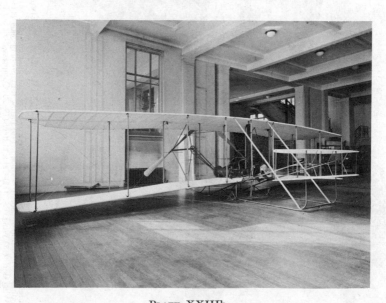

PLATE XXIIIb
The Wrights' aeroplane *Kitty Hawk*.
(Both these illustrations are reproduced by courtesy of the Director
of the Science Museum, South Kensington, London).

PₑCASTLₑ

Plate XXIV: German Flying Bomb (V.1)

(1) Air log. (2) Pressure plate operating electric fuse. (3) Wooden sphere containing compass. (4) Switch operating fuse. (5) Fuse. (6) Emergency firing device for fuse. (7) Impact fuse. (8) Second ditto. (9) Portion of 2000-lb warhead. (11) Warhead bolted to fuel tank. (12) Filler cap. (13) Lifting lug. (14) Fuel sump. (15) Main spar. (16) Wing skin. (17) Wing ribs. (18) Cable cutter. (19, 20) Compressed-air bottles. (21) Air pressure valve. (22) Fuel pump. (23) Pitot tube to 22, compensating for speed. (24) Battery. (25) Automatic pilot. (26) Mileage counter. (27) Rudder servo-motor. (28) Elevator servo-motor. (29) Lift spoiler for initiating dive. (30) Tube containing trailing aerial. (31) Nose cowling. (32) Shaped air duct. (33) Grill. (34) Venturi block. (35) Sparking plug for starting only. (36) Impulse duct engine. (37) Rear supporting fork. (Crown copyright. From a model in the Science Museum, South Kensington, by courtesy of the Director.)

work of machines is concerned, it will be necessary to reckon air-fleets not in hundreds, but in thousands.

(*The Aeroplane in War*. Claude Grahame-White and Harry
Harper. Werner Laurie. 1912.)

THIRTY-THREE YEARS AFTER

On August 6th 1945, shortly after 8 a.m., an American Super-Fortress flying at 30,000 feet, dropped a single atomic bomb over the Japanese mercantile city of Hiroshima. The bomb exploded over the city-centre. Three days later, on August 9th, just after 11 a.m., a Super-Fortress, flying at the same height, which had found its primary target cloud-obscured, dropped a second atomic bomb over the industrial city of Nagasaki. This bomb exploded over the city's factory area. In Hiroshima more than four square miles of city were destroyed and 80,000 people were killed. In the smaller city of Nagasaki about one and a half sq. miles were destroyed and nearly 40,000 people were killed. The causes of destruction and death differed in many points from those which had acted in the conventional raids of the past. It was clear that bombing had changed its character and its scale beyond recognition.

(*The effects of the Atomic Bombs at Hiroshima and Nagasaki.*
H.M.S.O. London.)

CHAPTER SEVENTEEN

The Atom in the Nineteenth Century

Developments of the Atomic theory

The eighth chapter of this book describes the atomic theory as proposed by Boyle and Newton. The men of that age saw that an atomic theory could be the way of explaining the world mathematically in terms of the positions, dimensions and masses of atoms and the forces exerted by them. It was one thing to propose this as a programme, but quite another to give it effect; and actually very little was done with the atomic theory until the beginning of the nineteenth century. In the nineteenth century the idea of atoms was used to explain a number of phenomena and predict new ones, but almost nothing was learnt concerning the behaviour of individual atoms as distinguished from that of atoms *en masse*: the investigation of the individual atom has been the principal and fruitful work of the twentieth century.

The explanation of crystals in terms of atoms

About the first successful use of the atomic theory was to explain the form of crystals—the bodies, bounded by plane faces, which are formed when a solid is formed from a liquid or a gas. Each chemical substance forms a crystal different from any other. Robert Hooke in 1665 had suggested that the form of some crystals could be explained if they consisted of tiny spheres regularly packed in contact with each other, but the idea lay dormant until near the end of the eighteenth century when the Abbé Haüy showed that all crystal forms could be built up of tiny blocks all of the same shape. (Plate XXVII.) It did not follow that these blocks were actual atoms or molecules (as indeed they are not) but the idea was a fruitful one. The crystallographers of the nineteenth century measured the angles of crystals and worked out the geometry of their forms, and, indeed, it proved that all crystals had forms that could result from packing together exactly similar blocks, but how these blocks (lattices) were related to the atoms and molecules did not appear till the year 1910, when v. Laue and Bragg made the discovery of the effect of crystals in diffracting X-rays, which has since given us the most exact knowledge of the arrangement of atoms and molecules in the solid state.

Explanation of chemistry in terms of atoms

Lavoisier was an atomist, but he did not attempt to discover the number and kind of atoms contained in the molecules of compounds. This problem was partly solved by John Dalton (between 1802 and 1808) though the full solution did not come for half a century. Dalton studied the proportions by weight in which elements combine to form compounds, and showed that these could be explained if every element was considered to consist of precisely similar atoms having a weight characteristic of that element, and if these atoms combined in simple numerical proportions. He assumed that the simplest compound of any two elements contained one atom of each and on this basis worked out the relative weights of the atoms of various elements. But Dalton's assumptions were not soundly based. One part by weight of hydrogen combined with eight* parts by weight of oxygen to form water: if the molecule of water was one atom of hydrogen combined with one atom of oxygen, then clearly the weights of these atoms were as one to eight. But did the molecule of water really consist of one atom of each of the two elements? That and similar problems exercised the minds of chemists for another fifty years. It was finally solved by adopting the principle that equal volumes of gases, under the same conditions, contain equal numbers of molecules. This principle, by arguments that cannot here be set out, led, from about 1860, to agreement on the number and kind of atoms in the molecule of any compound. This was expressed in the chemical formula of the compound. Dalton expressed formulæ by means of special symbols (Fig. 50) but within a few years Berzelius brought in the present method of denoting atoms by letters and the number

* The figure adopted by Dalton was seven.

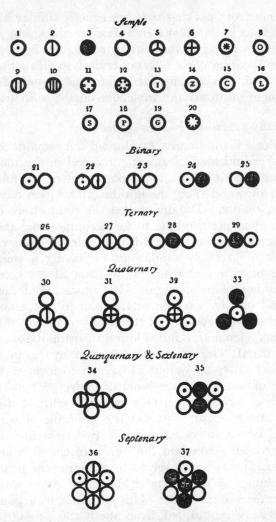

FIG. 50 —Dalton's Chemical Symbols. (1) Hydrogen, (2) Azote,
(3) Carbon, (4) Oxygen, (5) Phosphorus, (6) Sulphur, (7) Magnesia,
(8) Lime, (9) Soda, (10) Potash, (11) Strontia, (12) Baryta, (13) Iron,
(14) Zinc, (15) Copper, (16) Lead, (17) Silver, (18) Platinum, (19) Gold,
(20) Mercury, (21) Water, (22) Ammonia, (23) 'Nitrous Gas', (24)
Ethylene, (25) Carbon Monoxide, (26) Nitrous Oxide, (27) Nitric
Acid, (28) Carbon Dioxide, (29) Methane, (30) 'Oxynitric Acid',
(31) Sulphuric Acid, (32) Sulphuretted Hydrogen, (33) Alcohol, (34)
Nitrous Acid, (35) Acetic Acid, (36) Ammonium Nitrate, (37) Sugar.

of each present by subscript figures. Thus the formula C_2H_6O means 'a molecule consisting of two atoms of carbon, six of hydrogen and one of oxygen'.

Organic chemistry, concerned with the compounds of carbon, developed rapidly from 1830 onwards. Enormous numbers of carbon compounds were made. It was found that two different compounds might have the same formula, and so there developed the idea that not only were the number and kind of atoms in the molecule important, but also their grouping. The determination of the structure of these groupings was greatly aided by the idea of *valency*—that each atom had a certain combining power—that an atom of carbon could combine with 4 atoms of hydrogen, but an atom of nitrogen with only 3, and so forth. This combining power was represented in the formula by lines, denoting what were called *valency-bonds*. Thus, between 1800 and 1860, atoms were promoted from being merely particles, to being particles of fixed weight and combining-power. The idea of the molecule grew from that of a mere huddle of atoms, first, to an association of atoms in definite *proportions* and then to an association of a definite *number* and kind of atoms, and finally to a *structure* of atoms linked in a fixed and definite pattern. Thus benzene started as CH, later was characterised as C_6H_6, and finally (in 1865) set out as the hexagonal structure:—

But even at that time chemists were doubtful whether these patterns really existed or were just a useful convention for expressing chemical behaviour; so great interest was evoked when certain compounds were discovered whose formulæ could only be expressed by writing them three-dimensionally. Thus Louis Pasteur investigated two forms of sodium ammonium tartrate which differed only in their power of rotating the plane of polarised light and in the forms of their crystals, which differed as a left-boot differs from a right-boot (Fig. 51). Their formulæ were written as three-dimensional arrangements of atoms differing in the same way. The organic chemists thus gained very definite ideas about the reality of molecules, and what has since proved to be very correct ideas about their shapes. These ideas were the foundation of organic synthesis, some of the results of which are discussed in Chapter XVII.

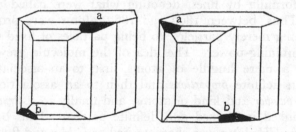

FIG. 51.—Enantiomorphic crystals of sodium ammonium *d*-tartrate and *l*-tartrate.

There was of course very much that the nineteenth-century chemists did not know about the molecule. They did not know its size, the distances between the atoms nor the extent of their relative motion, if any. All this remained to be demonstrated in the twentieth century.

The nature of the forces that bound the atoms into molecules was deduced from the facts of electrolysis (p. 247). Electricity could result from chemical combination (as in the voltaic batteries) and it could decompose some chemical compounds; accordingly Berzelius and others thought of the atoms as bound together by electrical forces; Faraday formulated definite laws connecting the qualities of substance undergoing chemical reactions and the quantity of electricity required to bring these reactions about. The idea of electrical forces binding the atoms together was fruitful in explaining the behaviour of acids, bases and salts, but not of other compounds, especially those in the great field of organic chemistry. Towards the end of the century, the idea was the basis of the ionic theory, p. 280, which however brought almost as many problems as it solved.

The kinetic approach

The chemists and crystallographers at first treated the atoms and molecules as if they were stationary. But as long ago as 1620, Bacon had suggested that heat was the motion of particles, and Rumford (p. 159) brought convincing evidence of this theory. This mere idea of motion was developed by a mathematical treatment into the Kinetic Theory. Gases were its first subject. It was shown that a great many of the properties of gases (such as Boyle's Law, Charles's Law, Avogadro's Law) and also a great many of the properties of heat could be deduced from the view (pp. 343-44) that gases consist of molecules in unceasing chaotic motion, at an average distance so great in proportion to their size that their volume and mutual attractions and repulsions were negligible. The molecules were treated as spherical, perfectly elastic, and impenetrable. The energy of these particles was the heat energy of the gas.

These assumptions were realised to be approximations; and their value was in representing a gas by a *model*, from which calculations could be made and then checked against the behaviour of real gases. This kinetic theory of gases was worked out in great detail between 1850 and 1870. It was found that no gas perfectly obeyed the simple gas-laws; but the investigation of the difference between the 'perfect gas' of the theory and the real gases of the laboratory led to valuable ideas about the volumes of the molecules and their attractions and repulsions. It was found that gases were most like the 'perfect gas' when at low pressures and high temperatures, but at high pressures and low temperatures, as they neared their points of liquefaction, they deviated from the gas laws.

This study led to the understanding of the theory of liquefaction of gases. Faraday, by simultaneous compressing and cooling gases, had succeeded, by 1844, in liquefying most of the common gases, but a few, e.g. hydrogen, oxygen and nitrogen, resisted all efforts to liquefy them by such means.

Oxygen was first liquefied in 1885, but the really successful method of liquefying the so-called permanent gases was not worked out until 1895, when the Joule-Thomson effect was utilised. The highly compressed gases were liberated through a valve and the expanded and cooled gas was used to cool the gas that had not yet expanded. This cold compressed gas when released, fell to an even lower temperature, and the temperature thus progressively fell until the gases liquefied. This work proved to be enormously valuable. The use of liquid air has proved invaluable both in physics and in chemistry; low temperature physics has progressed until temperatures only 0.003° C. above the absolute zero can now be attained.

Not only gases, but also liquids and solids, could be treated as assemblages of moving particles. The treatment of solids and liquids in this way was not at first so fruitful because their particles were very close together and the unknown forces between them played an important part in determining their behaviour. But solutions were admirably treated in this way from about 1870. It was shown that the dissolved substance behaved analogously to a gas, and so it became possible to apply the gas-laws to dissolved substances, and so to find out their molecular weights. This led to the discovery that the molecules of acids, bases and salts in solution are broken up into electrically charged particles called ions.

The kinetic theory of chemical reactions

The atomic theory threw a little light on the difficult question of what happened when a chemical reaction took place. Measurements showed that the speeds of chemical reactions differed according to the conditions of temperature and pressure (or concentration of a solution). According to the atomic and kinetic theory the molecules must collide in order that they should react, so the quantity of material reacting in a given time should be proportional to the number of collisions

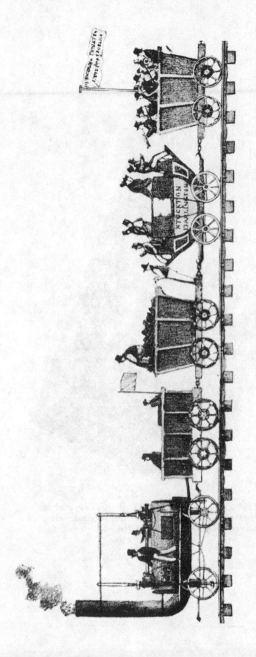

PLATE XXV

The opening of the Stockton and Darlington Railway in 1825

PLATE XXVI

Haüy's drawing of a scalenohedral crystal of
Iceland Spar, built up of rhomboidal units,
having the same form as the cleavage fragments of the spar.

between the two kinds of molecule concerned. This idea led to some valuable results, but it was apparent that other factors than numbers of collisions were concerned and Arrhenius in 1889 put forward the idea of *active molecules*. He supposed that, in order to react, a molecule had to acquire extra energy from some external source. This idea has been developed in recent years and we are coming near to being able to calculate the speed of the simplest reactions from the properties of the molecules taking part.

Thermodynamics

The atomic approach to such questions had the disadvantage that, in the nineteenth century (as even to-day) all too little was known about the individual atoms. Indeed in the eighteen-eighties some chemists began to have doubts as to whether these atoms were more than things of the mind. The school of Mach regarded atoms as mental artifices because they could not be perceived by the senses. It seems strange to refuse to admit the reality of anything smaller than the senses can perceive, and the tendency since c. 1900 has been increasingly to disregard the distinction between that which we perceive and that which we infer. This view concerning the hypothetical character of atoms favoured calculations based on measurements of the energy that a chemical reaction could give out, and making no assumptions concerning the mechanism of the change. Calculations of this kind became possible once the law of conservation of energy (p. 153) had been enunciated. It was not possible to calculate the speed of a reaction in this way, but it threw light on the final state of a chemical system after it had had time to react as completely as possible, and that without considering the behaviour of the individual atoms and molecules. Willard Gibbs and van't Hoff were the two great pioneers on this subject. Its practical result was to show conditions under which many unpromising reactions could give a yield that was industrially valuable. To it we owe the fixation of nitrogen, synthetic petrol, and a variety of important chemical industries.

Thus in the nineteenth century the idea of atoms was useful

and certain aspects of chemistry could have got nowhere without it. It was clear that matter behaved 'atomically', but the discovery of anything concerning the actual size, weight and other properties of individual atoms seemed almost hopeless. Yet from the beginning of the twentieth century there was a flood of evidence on this subject from very varied sources. This is considered in Chapter XIX; but we must first look at some of the practical consequences of the chemical ideas of the nineteenth century.

Extracts Concerning the Earlier Atomic Theory

HOOKE EXPLAINS CRYSTAL FORMS BY THE PACKING OF GLOBULAR PARTICLES

". . . so I think, had I time and opportunity, I could make probable, that all these regular Figures that are so conspicuously *various* and *curious*, and do so adorn and beautifie such multitudes of bodies, as I have above hinted, arise onely from three or four several positions or postures of Globular particles, and those the most plain, obvious, and necessary conjunctions of such figur'd particles that are possible, so that supposing such and such plain and obvious causes concurring the *co-agulating particles* must necessarily compose a body of such a determinate regular Figure, and no other; and this with as much necessity and obviousness as a fluid body encompast with a Heterogeneous fluid must be protruded into a Spherule of Globe. And this I have *ad oculum* demonstrated with a company of bullets, and some few other very simple bodies; so that there was not any regular Figure, which I have hitherto met withall, of any of those bodies that I have above named, that I could not with the composition of bullets or globules, and one or two other bodies, imitate, even almost by shaking them together. And thus for instance may be find that the *Globular* bullets will, of themselves, if put on an inclining plain, so that they may run together, naturally run into a *triangular* order, composing all the variety of figures that can be imagin'd to be made out of equilateral triangles; and such will you find, upon trial, all surfaces of *Alum* to be compos'd of: . . ."

(*Micrographia*, by Robert Hooke. 1665.)

DALTON ON ATOMIC WEIGHTS

In all chemical investigations it has justly been considered an important object to ascertain the relative *weights* of the simples which constitute a compound. But unfortunately the enquiry has terminated here; whereas from the relative weights in the mass, the relative weights of the ultimate particles or atoms of the bodies might have been inferred, from which their number and weight in various other compounds would appear, in order to assist and to guide future investigations and to correct their results. The following general rules may be adopted as guides in all our investigations respecting chemical synthes.

1st. When only one combination of two bodies can be obtained, it must be presumed to be a binary one, unless some cause appear to the contrary.

2nd. When two contributions are observed, they must be presumed to be a binary and ternary.

(There follow five more rules.)

From the application of these rules, to the chemical facts already well ascertained, we deduce the following conclusions; 1st. That water is a binary compound of hydrogen and oxygen, and the relative weights of the two elementary atoms are as 1 : 7, nearly; . . .

That carbonic oxide is a binary compound, consisting of one atom of charcoal, and one of oxygen together weighing nearly 12; that carbonic acid is a ternary compound . . . consisting of one atom of charcoal, and two of oxygen weighing 19; &c. &c.

(*A new system of Chemical Philosophy*. John Dalton. 1808. p. 215.)

CHAPTER EIGHTEEN

New Kinds of Matter

The Importance of Chemistry

At the end of the eighteenth century there were in use only a very few kinds of stuff which had not been known centuries before. Houses, tools, utensils, textiles, were all made of the same materials as of old—namely, wood, stone, bricks, mortar, the four common metals and their alloys, glass, and ceramics, together with the plant and animal products, such as cotton, linen, hemp, silk, wool, leather, etc. The eighteenth-century drugs and dyes were, with very few exceptions, natural plant or animal materials, some excellent, some very imperfect. In 1944 on the other hand we employ, not only in industry but in everyday life, an enormous variety of materials that had never existed before the labours of the chemist brought them into being. A glance round a modern house will quickly show the truth of this. Stainless steel, ferro-concrete, aluminium, enamelware, plastics, rayon, aspirin, cleaning-solvents, gas, paraffin, vulcanised rubber—these are a few of the common materials of life that did not exist in 1800.

Chemistry becomes useful

Chemistry had made more than a little progress by 1800. Quite a number of the elements had been discovered, and new compounds were being made or recognised. Moreover the growing industries were very ready to use anything that chemistry provided. Chloride of lime was quickly taken up for bleaching textiles, the chlorates for making matches, coal-gas for lighting, yet it must be said that, up to about 1850, industrial progress was much more in the direction of making the old materials into new mechanical devices than in making new materials.

The Metals

Up to 1850 the common metals and alloys in use were copper, brass, bronze, lead, tin, pewter and iron; while zinc was coming into use more widely. Three kinds of iron were used: (1) cast-iron; which was cheap but brittle, and was unsuited for anything which had to undergo tension of bending; (2) wrought-iron or soft-iron; this was a pure iron, smelted from the pure Swedish ores, or made from cast-iron by a laborious and therefore expensive process called *puddling*. (3) Steel, which could only be made in small quantities and by a lengthy process, and which was therefore too expensive for use than anything but tools and small machine-parts.

It was therefore a great event for industry when Bessemer in 1856 first discovered a way by which cast-iron could be cheaply and on a large scale turned into *mild steel*, which was strong and not brittle and was capable of being forged with the hammer, rolled, or turned on the lathe. This discovery led to all manner of engineering improvements—first of all a great lightening and cheapening of all manner of machinery, then the use of steel rails (p. 211), steel-skeleton buildings, and all kinds of constructional steelwork, ferro-concrete, etc. The Forth Bridge (opened 1890) is one of the first examples of constructional steel-work and could not have been constructed before mild steel became available.

One of the chief merits of the internal combustion engine (p. 264) was its lightness, and from the beginning of the twentieth century the production of motor-cars and aeroplanes created an urgent need for large quantities of a lighter metal. Aluminium had been known for many years, but had remained very expensive and little more than a chemical curiosity. When cheap electricity had become available, it was possible to make aluminium at a low cost, and from about 1900 the production of this metal increased enormously. First it was used for machine-parts, then later for kitchen utensils and many small light objects. For these latter purposes, however, it shows signs of giving way to plastics and stainless steel.

The production of special steels—i.e. alloys of iron, carbon and one or two metals such as manganese, chromium, vanadium, or tungsten, etc. (which import such qualities as extreme hardness or toughness), began in the eighteen-seventies, but became really important only after 1900. These steels have added to the strength, compactness, and output of machinery; but stainless steel, which has most affected domestic life, was patented as lately as 1916. Stainless steel contains about an eighth of its weight of chromium. The making of chromium (and of most of the rarer metals) requires the electric furnace, or the use of aluminium, made by means of electricity; so stainless steel is one of the very numerous modern inventions that was made possible by that most crucial invention of the efficient dynamo (p. 249).

Chemistry and Food

The chemists, during the nineteenth century, worked out the requirements of green plants, which are the ultimate source of all our food. They showed that their main substance was derived from air and water, but that they took from the soil small quantities of absolutely necessary inorganic salts, notably nitrates and phosphates. Biologists worked out the manner in which bacteria convert the nitrogen and phosphorus compounds contained in plant and animal remains into nitrates and phosphates, which other plants can feed upon.

It thus became apparent that the crops, taken from the earth and consumed by man, were steadily depleting the soil of the substances necessary for plant-life, and the practice of replacing these by adding inorganic fertilisers became general at the end of the century. Chemical industry after 1900 found means of converting the nitrogen of the air into nitrogen compounds, and after 1920 these became the most important of fertilisers. Rock-phosphate has been extensively mined and made into super-phosphate: basic slag, a by-product of iron-smelting, is likewise a source of phosphates. These fertilisers, new inventions of the chemists, are in fact a principal source of our food; without them, it is doubtful whether the world-population could be maintained, let alone further increased.

Organic Synthesis

Lavoisier said: "La vie est une fonction chimique", and it is true that the characteristic activity of our bodily part consists of the chemical reactions of complicated carbon compounds. It follows then that Organic Chemistry—the chemistry of carbon compounds—is the approach, first, to making those materials that we obtain from animals and plants, and substitutes for or improvements on them: secondly, to the preparation of curative drugs, substances that will enter into chemical reaction with our bodies and influence them.

Organic chemistry was not at all well understood until the eighteen-sixties when the rules of combination (valency-relationships) between carbon, hydrogen, oxygen, and nitrogen atoms were made clear by Frankland, Kekulé and others. From this time the discovery of new classes of compounds went ahead with gigantic speed and the great fine-chemical industry was built up, chiefly in Germany.

Dyes and Drugs

Almost all dyes and most of the effective drugs known to man before the eighteen-fifties were prepared from plants or occasionally animals. Their molecules were known to be compounded of atoms of carbon, hydrogen, oxygen and nitrogen, but our knowledge of the structure or pattern formed by these atoms remained very imperfect, as also did the technique of putting them together to form molecules of the patterns we desired. In 1856 W. H. Perkin, as a youth of eighteen, conjectured that quinine might be made by the chemical reaction of substances whose molecules contained the right number and kind of atoms to make the quinine molecule. We know to-day that such a result requires not only the right kind and number of atoms but their arrangement in the right pattern; and indeed quinine has never yet been synthesised. The substances that Perkin used in his attempt, actually formed a dirty-looking slime, the investigation of which led to the discovery not of synthetic quinine, but of *mauve*, the first synthetic dye (p. 295). Since the composition of dyes was not understood, not many new ones were discovered until the eighteen-seventies, when the theory of organic chemistry began to be well developed. Then they were discovered first by dozens, then by hundreds, and brilliant and lovely shades, hitherto unknown save in the petals of flowers, could be dyed on silk and wool, and later on cotton. This was in itself an important advance, for colour, though not a necessity, is a delightful amenity of human life; but the rise of the dye manufacture was something more—namely the means of making an *organic chemical industry*. This has continued to make spectacular progress in the production of fast and varied dyes, but it has also created perhaps the most valuable industry in the world, that of synthetic drugs.

Synthetic drugs

Organic chemistry at a very early date produced some simple compounds which yet contributed incalculably to human happiness. Ether and chloroform, brought into use as anæsthetics in 1846–7, though discovered earlier, are the prime examples, for they not only removed pain from surgery, but made surgery acceptable both for the medical man who was unwilling to inflict terrible pain and for the patient who would rather be ill than endure it. After Lister's work (p. 194) the organic chemists provided the carbolic acid, discovered in 1834, which enabled antiseptic surgery to be performed, and which in time abolished the fearful maternal mortality, which, in some hospitals, rose to one death in four. After 1880 when the bacterial theory of disease (pp. 195*ff*) was adopted, dozens of new disinfectants, with advantages over carbolic acid, were synthesised by the chemist, and this has culminated in the modern non-poisonous and non-irritating synthetic antiseptics. Attempt to synthesise drugs for internal use continued. Chloral, a good hypnotic, was one of the first, in the 'eighties and 'nineties came antipyrin, phenacetin, acetylsalicyclic acid (aspirin), sulphonal and many others, which, while they did not *cure* any disease, were useful to alleviate symptoms—to lower the temperature relieve pain, and the like.

The pharmacologist dreamed that he might find a drug which, when administered to a patient suffering from a germ-disease, would kill the germs without harming him; but such drugs proved elusive until the twentieth century. In 1907 Paul Ehrlich discovered *trypan red*, which operated in this way upon organisms, akin to the sleeping-sickness parasite, which caused African cattle-plagues, but this was not a success with man; and in 1910 he made the great discovery of '606' or *salvarsan* which, in improved forms, has enormously reduced the ravages of syphilis, a venereal disease whose terrible late results filled a considerable part of the nineteenth-century medical text-books. The latest and most spectacular result of organic chemistry has been the synthesis of the sulphonamide drugs (p. 198) which since 1935 have revolutionised the treatment of septic diseases, pneumonia, meningitis, gonorrhœa, and many conditions which were formerly fatal.

The synthesis of substances than the body needs is another recent development. The vitamins—those oddly assorted materials of which traces are required in our food in order that health should be maintained—can now for the most part be synthesised, as also can many of the potent substances secreted by the ductless glands. Here we are drawing near to the chemistry of the body's own processes. It is difficult to estimate what degree of control ovei the body's workings may be attained by the use of that knowledge of the chemical processes of the cell, which we are just beginning to learn (p. 312), but there is little doubt that we are on the threshold of biochemical discoveries relative to the control and modification of human bodily functions.

Organic materials

Wood, leather, horn, paper, textiles, etc., are organic materials that have been used from antiquity for making objects of use and articles of clothing. Their qualities of lightness, strength, warmth, flexibility, etc., marked them off from all mineral products, but until the twentieth century we had very little knowledge of why they possessed these valuable properties. But even without this knowledge some valuable additions to them were made. The art of vulcanising rubber with sulphur dates from c. 1840, and gave the world its first waterproof elastic material. The use of more sulphur gave vulcanite, one of the first *plastics*, that is to say materials that can be moulded in a hot press and which will harden on cooling. Pressure-moulding is a very cheap and convenient way of making irregularly shaped objects; so a new plastic, celluloid, a mixture of nitrocellulose and camphor, was greatly welcomed in the eighteen-eighties. It was the first flexible transparent material and so it soon led to the photographic film and the motion picture, previously impossible. It is strange to think how the discovery of nitrocellulose in 1846 was the beginning of four great industries: high-explosive, the hand-camera, the motion-picture, and artificial silk. Celluloid was a very convenient material; but it was exceedingly and dangerously inflammable. In 1897 came a new step, the use of the non-inflammable *casein*, which is a horny substance that is contained in skim-milk and gives its firmness to cheese. Mixtures of casein and pigments could be moulded, and formed a very good substitute for ivory. In 1909 came the really great advance. Leo Baekeland, who was trying to make a substitute for the resins used in varnish manufacture, by mixing solutions of phenol (carbolic acid) and formaldehyde made a resinous substance which, like celluloid or vulcanite, could be moulded into any shape in a hot press, but, unlike them, set hard and thereafter could no more be softened by heat. This new substance, *bakelite*, was developed but slowly, but after the first world-war it came into use for all manner of small articles and especially electrical accessories. Bakelite was only obtainable in dark colours; but in 1924 Pollak discovered light-coloured plastics made from urea and formaldehyde. From 1935 beautiful and strong "organic glasses" began to be made. Meanwhile new uses began to be found for plastics, production hugely accelerated, and between 1935 and 1940 the quantity produced was quadrupled. Plastics (usually combined with some natural fibrous material such as wood-flour or paper) are now the material of choice where light, strong, and accurately-finished articles are required in quantity. Weight for weight, plastic materials are more expensive than wood, glass or metals, but they are moulded into shape with such economy of labour and so little waste of material that their original cost is in many cases not an important factor.

Plastics will obviously not displace the older materials entirely or even largely, because they are not hard like glass or china, nor are they heat-proof like metals. They do, however, form the first new class of constructional materials that have been brought into use since the time of ancient Rome; and they initiate a new era in which materials will be designed for the object that is to be made, instead of the design of the object being adapted to the material.

Examples Illustrating Organic Chemical Synthesis

THE STORY OF CHLOROFORM

I. *J. Liebig in* 1831 *discovers chloral and by treating it with alkali obtains a "chloride of carbon"—actually chloroform.*

(1) Dry chlorine is passed through hot absolute alcohol, until no more hydrochloric acid is evolved; the alcohol then solidifies to a white crystalline mass.

(2) These crystals consist of a peculiar body, which I call Chloral (from *Chlor* and *Alkohol*), in combination with water. . . .

(5) By the action of solutions of alkalis in water Chloral is converted into a new chloride of carbon and formic acid.

(6) The new chloride of carbon can also be easily prepared by distillation of alcohol with an excess of chloride of lime.

(Ueber der Zersetzung des Alkohols durch Chlor. von J. Liebig. *Poggendorf's Annalen*, 23.4.44.)

II. *M. E. Soubeiran, in the same year, discovers chloroform independently.*

. . . To assure myself of this, I mixed a very concentrated solution of chloride of lime with alcohol; the mixture was heated and an odour of chlorine became noticeable; on bringing it to the boil, a large quantity of a white precipitate was formed and there distilled over a liquid with a very pleasant smell and a sweet taste.

(Soubeiran. *Annales de Chimie et de Physique* [2] 48, 131.)

Soubeiran, unlike Liebig, realised and proved that the new liquid contained not only carbon and chlorine but also hydrogen; but both he and Liebig arrived at incorrect formulæ. Liebig's formula was C_4Cl_5; Soubeiran's was CH_2Cl_2; but Dumas analysed chloroform in 1843, determined its density, aud made its formula to be $C_4H_2Cl_6$, which using modern values of atomic weights (carbon = 12, not 6, as then supposed) leads to the formula $CHCl_3$, adopted to-day.

III. *The need for an anæsthetic*

Professor Miller recalls the olden days when a patient, in a panic of fear, was being removed from the ward: 'His progress might be traced by frightful yellings, or, at least, by sobs of deep distress, and occasionally a number of stout assistants scarcely sufficed to prevent on the way a self-effected rescue and escape. All this was bad, painful, injurious, unseemly. All such scenes are now unknown.' To lead people to use chloroform, Professor Simpson received a letter from Dr. George Wilson, who had undergone an operation before the days of anæsthesia, in which he describes his anxiety of mind thinking of what lay before him, what he suffered by the surgeon's knife, till he saw 'the bloody, dismembered limb lying on the floor'. 'The particular pangs are now forgotten,' he wrote, 'but the black whirlwind of emotion, the horror of great darkness, the sense of desertion by God and man, bordering close upon despair, which swept through my mind and overwhelmed my heart, I can never forget, however gladly I would do so. Even now, if by some Lethean draught I could erase the remembrances of that time, I would drink it.'

(Those who think we would be better without science must consider this passage.)

IV. *Sir James Y. Simpson experiments on himself to find an anæsthetic drug which would be better than Ether, used the year before in America.*

'Latterly, in order to avoid, if possible, some of the inconveniences pertaining to sulphuric ether,* particularly its dis-

---
* i.e. ordinary or ethyl ether.

agreeable and persistent smell, its occasional tendency to irritation of the bronchi during its first inspirations, and the large quantity of it required to be used (more especially in protracted cases of labour), I have tried upon myself and others the inhalation of different other volatile fluids, with the hope that some one of them might be found to possess the advantages of ether without its disadvantages. For this purpose I selected for experiment, and have inhaled, several chemical liquids, of a more fragrant or agreeable odour, such as chloride of hydrocarbon, etc. Then Mr. Waldie, a Linlithgowshire man, first named perchloride of formyle* as worthy among others of a trial.'

' . . . I am sure you will be delighted to see part of the good results of our hasty conversation. I had the chloroform for several days in the house before trying it, as, after seeing it such a heavy, unvolatile-like liquid, I despaired of it, and went on dreaming about others. The first night we took it, Dr. Duncan, Dr. Keith, and I all tried it simultaneously, and were all "under the table" in a minute or two.'

(Both III and IV are from *Sir James Y. Simpson*, by E. Blantyre Simpson. Edin. & Lond. 1896.)

V. *The first operation under chloroform.* [Chloroform was first used in midwifery, but within a few months it was tried out in surgical practice.]

"I append notes, obligingly furnished to me by Professor Miller and Dr. Duncan, of the three cases of operation. The first two cases were operated on by Professor Miller, the third by Dr. Duncan. In applying the chloroform in the first case, I used a pocket-handkerchief as the inhaling instrument; in the last two I employed a hollow sponge.

Case I.—'A boy, four or five years old, with necrosis † of one of the bones of the fore-arm. Could speak nothing but Gaelic. No means, consequently, of explaining to him what he was required to do. On holding a handkerchief, on which some chloroform had been sprinkled, to his face, he became frightened

* Perchloride of formyle was an alternative chemical term for chloroform.

† Death of a portion of the bone with consequent inflammation and seps:s of the surrounding part. The partially separated dead bone (sequestrum) had to be removed before healing could take place.

and wrestled to be away. He was held gently, however, by Dr. Simpson, and obliged to inhale. After a few inspirations he ceased to cry or move, and fell into a sound snoring sleep. A deep incision was now made down to the diseased bone; and by the use of the forceps, nearly the whole of the radius, in the state of sequestrum, was extracted. During this operation, and the subsequent examination of the wound by the finger, not the slightest evidence of the suffering of pain was given. He still slept on soundly, and was carried back to his ward in that state. Half an hour afterwards, he was found in bed, like a child newly awakened from a refreshing sleep, with a clear merry eye, and placid expression of countenance, wholly unlike what is found to obtain after ordinary etherization. On being questioned by a Gaelic interpreter, who was found among the students, he stated that he had never felt any pain, and that he felt none now. On being shown his wounded arm he looked much surprised, but neither cried nor otherwise expressed the slightest alarm.

(Account of a New Anæsthetic Agent, as a Substitute for Sulphuric Ether, in Surgery and Midwifery. Sir J. Y. Simpson, Edinburgh. 1848.)

ST. JEROME DESCRIBES VITAMIN-A DEFICIENCY AND ITS CURE (A.D. 392)

From his thirtieth to his thirty-fifth year, (St. Hilarion) had for food six ounces of barley-bread, and vegetables slightly cooked without oil. But finding his eyes growing dim* and his whole body drawn up by a sort of scabbiness and dry eruption† he added oil‡ to his former food and up to the sixty-third year of his life followed this temperate course, tasting neither fruit nor pulse, nor anything whatsoever beside.

(*Life of Saint Hilarion*, § 4. St. Jerome. Translated in the *Nicene and Post-Nicene Fathers*. Vol. VI, *St. Jerome*, p. 305. Oxford, 1893.)

* Night-blindness or xerophthalmia.

† Follicular keratosis.

‡ Both these symptoms are occasioned by lack of the oil-soluble vitamin A or the provitamin A which can give rise to it. We do not know what oil the Saint ate, but many vegetable oils contain much provitamin A.

PERKIN AS A BOY OF EIGHTEEN MAKES THE
FIRST SYNTHETIC DYE-STUFF

At this period much interest was taken in the artificial formation of natural organic substances but . . . very little was known of the internal structure of compounds and the conceptions as to the method by which one compound might be formed from another were necessarily very crude. . . .

. . . As a young chemist I was ambitious enough to wish to work on this subject of the artificial formation of natural organic compounds. Probably from reading the above remarks on the importance of forming quinine, I began to think how it might be accomplished, and was led by the then popular additive and subtractive method to the idea that it might be formed from toluidine by first adding to its composition C_3H_4 by substituting allyl for hydrogen, thus forming allyltoluidine, and then removing two hydrogen atoms and adding two atoms of oxygen, thus

$$2 \underbrace{(C_{10}H_{13}N)}_{\text{Allyltoluidine,}} + 3 \; O = \underbrace{C_{20}H_{24}N_2O_2}_{\text{Quinine}} + H_2O$$

The Allyltoluidine, having been prepared by the action of allyl iodide on toluidine, was converted into a salt and treated with potassium dichromate; no quinine was formed, but only a dirty reddish-brown precipitate. Unpromising though the result was, I was interested in the action, and thought it desirable to treat a more simple base in the same manner. Aniline was selected and its sulphate was treated with potassium dichromate; in this instance a black precipitate was obtained, and, on examination, the precipitate was found to contain the colouring matter since so well known as *aniline purple* or *mauve*, and by a number of other names. All these experiments were made during the Easter vacation of 1856 in my rough laboratory at home. Very soon after the discovery of this colouring matter, I found that it had the properties of a dye and that it resisted the action of light remarkably well."

[Perkin, despite his youth and total inexperience, equipped

a chemical factory and made a great success of it. His heart, however, was in research, and eighteen years after, when he was thirty-six, he disposed of the business and returned to pure research, wherein he distinguished himself highly.]

(W. H. Perkin. *Hofmann Memorial Lecture.* Transactions of the Chemical Society. 1896. pp. 603–4.)

PAUL EHRLICH'S DELIBERATE SEARCH FOR A CURATIVE DRUG

The discovery which brought Ehrlich's name before the public was the introduction of salvarsan. It was no haphazard discovery, but it was based on a long series of investigations proceeding along well-defined lines. A necessary corollary of the side-chain theory* was the idea that a living cell, be it a cell of the body itself or of an invading microbe, was susceptible of certain chemical combinations, and therefore, if it were to be affected by any drug, that drug must possess certain chemical attributes or no effect would be produced. Hence it follows that definite chemical groups in the drug employed will have definite chemical actions, so that, were our knowledge sufficiently wide we should be able to predict the action of any drug before it was employed; and although our learning at present falls far short of this extent, yet we can gauge much of the action of an untried chemical substance if its constitution is known. From this basis Ehrlich started on an investigation of a cure for syphilis. He aimed at some substance which should destroy completely all the *spirochæta pallidæ*† in the body. He started with the fact that arsenic had been used with success in the treatment of syphilis and he designed to obtain a substance which, while causing a minimum of harm to the human body, should be a potent agent in destroying the spirochæta. He gave directions to those chemists who assisted him to prepare certain chemical substances,‡ thus reversing the

* Which involves the notion, that drugs act on tissues whether of bacteria or of their hosts by chemical combination with certain 'side-chains' or groups of atoms present in the molecules of the tissues on which it acts.

† Name then given to the germs of syphilis, discovered in 1905.

‡ i.e. to synthesize certain substances previously unknown.

usual process, where the chemist invents chemical substances which he offers to the physician for trial. Here the physician gave the direction to the chemical investigations, although of course, from the imperfect state of present knowledge the result of the substance could not be predicted with absolute certainty. After a large amount of work Ehrlich at length evolved salvarsan or dioxydiamino-arsenobenzol, "606" as it was called in the convenient phraseology of the laboratory (because it was the 606th preparation tested), and the drug was found at once to exert a wonderful influence upon syphilis. . . . With the results that have been obtained with salvarsan Ehrlich was not content; he continued his researches, and one result of these is neosalvarsan, which in some respects is an improvement on its predecessor. But Ehrlich, with the spirit of the true philosopher, continued looking for some substance which should be capable of producing a "sterilisatio magna",* and the work which he has not lived to accomplish remains to be carried out by others. . . . And (his work) will prove to be the forerunner of many therapeutic triumphs, for it is the outcome of practical work along lines that can be reproduced and extended.

(Obituary Notice of Paul Ehrlich, *Lancet*, August 8, 1915.)

---

* i.e. a substance which by a few doses will destroy the whole of the parasites of a specified kind in the tissues. The sulphonamide drugs come near to this, though their action is not so much to poison bacteria as to check their multiplication while the body's natural defences are destroying them.

CHAPTER NINETEEN

The Atom in the Twentieth Century

Knowledge of the individual atom

The nineteenth century had used to great profit the idea that matter behaves atomically, but it could give scarcely any information as to the properties that belonged to the individual atoms. In the last fifty years that lack has been largely remedied. We no longer think of atoms simply as 'minute spheres' but as objects of known size and mass with structures concerning which a great deal is known. This knowledge relates

(1) to their numbers, masses and dimensions,
(2) to their internal constitution,
(3) to their laws of motion,
(4) to their formation and decomposition.

The numbers, masses, and sizes of atoms and molecules.

The problems of counting the atoms in a measurable quantity of matter was seriously attacked in the years 1900–1910, and a surprising number of ways of doing it were found.

The earliest approach to detecting the activity of a single atom was by the study of the Brownian movement. J. Brown in 1827, just after the invention of the achromatic microscope lens had allowed high powers to be used, noticed that a great variety of minute particles, when suspended in water, were seen to be perpetually trembling (pp. 306–307). In the period 1870–1890 it was realised that this trembling was the result of the irregular hail of blows dealt to the suspended particles by the darting molecules of the surrounding liquid. In the early twentieth century it became clear that the energy of motion of such a particle must be the same as that of a molecule of the liquid, and three different ways of calculating from the Brownian movement the weight of a molecule and the numbers of them in a given weight of matter were worked out. As is shown by the table on page 309 a large number of other ways of calculating these quantities were found, and were in close agreement. Perhaps the best of these was the measuring of e, the charge on an electron, by the method of Millikan. We know the total charge on, say, a gram of chlorine ions, (*i.e.* chlorine atoms each having one electron more than normal), for this is the quantity of electricity needed to liberate a gram of chlorine by electrolysis. This charge divided by e would give the number of atoms in a gram of chlorine. This measurement of e was done by giving a small electrical charge to some of the droplets in a fine mist of oil and estimating their charge by the speed with which they moved under a strong electrical attraction. The charges found did not vary continuously; but if the least charge found was one unit, other oil drops had precisely 2, 3, 4 or more units. It was taken then that this unit charge was that of an oil-droplet that had but one electron attached to it. The result was that the number of atoms in a gram of hydrogen came out at 60,630,000,000,000,000,000,000,000 or 6.06×10^{23} and the weight of each of them to be $\dfrac{1}{6.06 \times 10^{23}}$ grams. Six or seven other ways of arriving at the result gave about the same figure. This experiment did not tell us the size of an atom. Its diameter had been estimated as about a ten-millionth of a millimetre from considerations of the thinness of the thinnest films of liquids and solids; and that estimate proved to be fairly close. The method of the analysing the structures of crystals, by passing through them an X-ray beam, which is reflected by different layers of atoms, was discovered by v. Laue in 1911 and simplified and developed by Bragg. It enabled the positions of atoms in a crystal to be estimated, and so gave a close idea of the amount of territory occupied by an atom. Atoms of different elements differ in diameter, but vary only from about one to about four ten-millionths of a millimetre.

Bragg's method proved enormously powerful. In its later developments it gives the exact structure of any crystal and enables the positions of all the atoms in certain molecules to be mapped. It gives, in fact, a way of discovering chemical formulæ by physical methods. The method is a very slow one, unfortunately, owing to the laborious calculations required, but the new calculating machines, such as the so-called 'electronic brain', will probably much diminish this difficulty.

We know now the size and weight of atoms with some exactness and can map their positions in a molecule or crystal. No very high degree of precision however is possible in estimating sizes of atoms, because, as we shall see, an atom has no boundary.

The structure of Atoms

The beginning of the idea that atoms had a structure may be thought to date from Mendeléeff's great discovery of the periodic law. He arranged the elements in order of atomic weight and showed that similar elements occurred *at regular intervals* in the list. This showed that there was some kind of relationship between the elements which had previously been thought of as being, as it were, isolated and arbitrary creations. The elements thus fell into real natural classes or families. The existence of gaps in some of these families led to the prediction of the existence of elements as yet unknown. The success of these predictions and the chemical discoveries that resulted from investigating the resemblances between the members of each family attracted great attention. None the less, there seemed to be no way of finding out what lay behind this periodic principle which linked together every kind of matter. It obviously represented something fundamental, yet represented it inexactly, because there were a number of minor differences between Mendeléeff's predictions and the actual facts.

The discovery of the meaning of Mendeléeff's classification came from physics rather than chemistry. We have already chronicled the discovery of the electron as a common constituent of all matter (p. 258). Electrons are negative electricity, so matter (which is electrically neutral) evidently contained positive electricity. The idea of the atom as containing electrons dates from 1897.

Meanwhile radioactivity was discovered and in 1902 Ernest Rutherford and Frederick Soddy showed that it was a disintegration of atoms with ejection of positive α-particles (helium nuclei), β-particles (electrons) and γ-rays, resembling X-rays. This was the first evidence of the transmutation of elements. Soddy and others independently arrived at the displacement law, which showed that as a result of the emission of these particles an element changed into another in a fully predictable manner. Soddy also announced the discovery of isotopes, elements of apparently identical properties but different atomic weights. Radioactivity was not only a great addition to our knowledge but a useful laboratory tool. It gave a source of high-speed particles, whose tracks could be registered by the Wilson cloud-chamber or the photographic plate. Rutherford in 1911 discovered that when α-particles were projected through metal foil, most of them were but slightly deflected, but a small proportion were violently deflected or even turned back on their tracks. This showed that a very small part of the space within the atom was occupied by a minute positive particle, much heavier than the electron. This was termed the nucleus of the atom. The nucleus of the lightest atom, that of hydrogen, he took to be an ultimate particle and named it the *proton*. Thus from 1911 we had evidence that the atom was a cloud of electrons around a minute heavy positive nucleus. But why were not the electrons attracted into the nucleus? This question could not yet be answered.

Meanwhile Moseley discovered that the number of the place of an element in the periodic table (its atomic number, N) was equal to the charge on the nucleus and so we could say that the nucleus of an atom had an electrical charge of N units* and was surrounded by N electrons. The elements therefore formed a *series*, the atom of each element having one more electron in its outer cloud than that of the element before it in the series.

The problem of the arrangement of these electrons was solved by Niels Bohr, who applied the quantum theory (p. 330) of Planck to the electrons, supposing that they could only have certain fixed energies corresponding to so many quanta of energy. The diameter and eccentricity of the orbit in which

* The unit is, of course, e, the charge of one electron.

the electron was supposed to move, depended on this energy. When an electron moved from a larger orbit to a smaller the balance of energy was given out as a quantum of radiation. This theory gave a wonderfully exact and detailed explanation of the spectra of the elements and so guaranteed itself as being near the truth.

The announcement of Bohr's theory was followed by twenty years of intensive work. In 1932, Chadwick's discovery of the neutron, a particle with the same mass as a proton but no electric charge, brought our knowledge of the structure of the atom to something like its present state. Thus we now regard the atom of an element with atomic number N and atomic weight W as having N electrons in the outer cloud, and in the nucleus N protons and W – N neutrons. The structure and inner workings of the nucleus remain largely unknown and it will be the work of the next generation of physicists to produce a theory of them that will explain the nuclear phenomena that are to-day of such intense importance. Around this nucleus are groups of electrons each specified by four quantum-numbers: the outermost group (and sometimes one or two of the inner groups) being capable of losing, gaining or sharing electrons. The losing or gaining of electrons makes an atom into an ion: the sharing of electrons by two atoms constitutes the valency-bond that links them into a chemical compound. Our knowledge concerning the atom enables us to include spectra, radioactivity, the periodic law, chemical combination, magnetic phenomena and many others in a single system—the greatest unification of science since the Newtonian theory.

The dynamics of atoms

The laws of dynamics were worked out on ordinary matter, —aggregates of countless billions of billions of atoms. They no more predict the behaviour of single atoms, than the Registrar's marriage statistics can predict the date of the marriage of a new born baby. All our everyday commonsense experience is of large-scale matter and so the laws that have been found to express the behaviour of minute particles, quanta of radiation etc. seem to us paradoxical. In order to make calculations and predictions about these, a new branch of physics, quantum-mechanics, had to be evolved. The mathematics it uses are formidable to the chemist of the old school: suffice it to say that it deals largely with the *probabilities* of finding particles in certain positions: an electron is not placed here or there, but is considered as all over the atom with a higher probability of being in some parts of it than others. This new way of considering matter is perhaps the greatest departure from the science of earlier centuries. The atomic theory was adopted, just because it gave a way of visualising every kind of change in terms of motions of atoms: to-day quantum mechanics entirely rejects visualisation and works on the purely intellectual level of interpreting mathematical formulæ which have been empirically found to be true. The method has justified itself by predicting and explaining numerous phenomena of which 'classical' dynamics could tell us nothing. In Chapter XXI more will be said concerning this new way of looking at the world.

Transmutation

The alchemists believed in the possibility of transmutation of metals because they saw no reason why these bodies should be exempt from the changes they saw in everything else; and also, because they had a strong desire to change base metals into gold. The men of the nineteenth century did not believe such changes possible because none of the processes they had applied to any element had been successful in changing it into another. The men of the twentieth century have discovered that transmutation had always been going on in nature and have accomplished it by artificial means, with the most startling results.

We have seen that Rutherford and Soddy, in 1902, showed that radioactive elements were slowly or rapidly transmuted into others and were giving out in this process vastly more energy per gram of matter than was given out in any other known process. Here was evidence that the atom was a storehouse of relatively vast quantities of energy. It was not possible, however, to influence this process in any way. We have also seen how at a later date (p. 235) the æonial continuance of radiation from the stars was explained by the theory that in them lighter elements were being transmuted into heavier with loss of mass, Einstein, in 1905, having showed that mass and energy were theoretically interconvertible (p. 335).

In 1919, the first artificial transmutation was demonstrated. The Wilson cloud-chamber gives us a way of photographing the tracks of high-speed particles through air. Rutherford photographed the tracks of α-particles, from radium, shooting through an atmosphere of nitrogen. A forked trail showed him that a nitrogen atom had swallowed up an α-particle and ejected a proton, thus being transformed into an isotope of oxygen. Many such transmutations of single atoms were recorded. Of course no weighable quantity of any element was produced, but the interest was in seeing how the nucleus responded to bombardment. It did not break up, but swallowed up the bombarding particle and shot out something quite different.

It was by bombarding the metal beryllium with α-particles that Chadwick discovered the neutron, and bombardment with neutrons proved to be much more effective than bombardment with α-particles, protons or electrons. In 1934 Madame I. Curie and F. Joliot made a further advance. They showed that some elements when bombarded became radioactive. Their nuclei swallowed up the bombarding particle and formed a new nucleus which, after hours, days, or months, ejected particles in just the same way as the natural radioactive elements. These artificial radioelements could be produced in minute but yet useful quantities by bombarding matter with streams of high-speed particles, accelerated by electric fields and guided by magnetic fields in such instruments as the cyclotron.

In 1938 came the portentous discovery of nuclear fission. The element uranium was bombarded with neutrons, not with the expectation of splitting the atom but rather of making new elements with heavier atoms than that of uranium, till then the heaviest known. But a new kind of decomposition took place. The uranium atoms did not always simply absorb the neutron, but some of them split up into two roughly equal parts and shot out several neutrons of great energy. It was even then apparent that this process could lead to an atomic explosion, for each neutron shot out might break up another atom, which would shoot out further neutrons to break up further atoms, and so on till the whole mass of atoms had taken part in the process.

The problem was, then, to discover why uranium did not explode in this way,—and to make it do so. The war of 1939–45 provided the stimulus. On both sides research was carried on with an expenditure of material resources unthinkable in peace-time. Air-attack prevented the Germans from diverting sufficient industrial resources to the task, and the scientists of the Allies, working in Canada and the U.S.A., succeeded. The reason why natural uranium did not explode was found to lie in the fact that it was a mixture of a small amount of the iso-tope U-235, which could undergo fission and so explode, with a large amount of U-238, which could not undergo fission and moreover absorbed the neutrons produced by the splitting of the U-235 atoms. So the next task was the separation of U-235 from U-238, which was known to be possible in theory, but extremely difficult. By the design of special plant, it was accom-plished. At the same time a hitherto unknown element, plutonium, capable of undergoing nuclear fission, was made from uranium in the famous pile,—an entirely new departure. The principles are beyond the scope of such a book as this; rather are we to examine the implications of our success in making material capable of nuclear fissions and the threat or promise of the future.

The fact of an unexampled weapon being placed in the hands of men is the first and present result. The atomic bomb attached to a high-speed projectile of rocket type can wipe out most of a large city and will prove very difficult to intercept. At present the somewhat frail hope is that, as with gas-attacks in the war of 1939–45, no one will care to begin a bout of atomic bombing. Large quantities of intensely dangerous radioactive material is produced in the making of the bomb and this may also be used as a weapon. A crop of human monstrosities, resulting from the action of the rays from atomic bombs upon the germ-cells of those exposed to them, is another possibility.

The second result is valuable: the ability to produce radioactive varieties of common elements. These have already done work as tracers: thus if a specimen of a drug made from radioactive material is administered to an animal the radioactivity will show to what organ it goes or in what form it is excreted. This is already advancing our knowledge of the working of the body and, therefore, our power to cure its ills.

The gigantic quantities of energy given out in the atomic pile promise to revolutionise engineering, but it will be a decade before we know how it will be utilised. Costs matter in peace in a way they do not in war. Numberless vistas have been opened up, e.g. the possibility of controlling the evolution of atomic energy so as to reach enormous temperatures without actual explosion: the possibility of using the energy of the fission of uranium to start similar processes that might continue on their own. We stand at the threshold of a new age of science.

Extracts Concerning the Individual Atom

BROWN DISCOVERS 'ACTIVE MOLECULES'

. . . "The plant was *Clarckia pulchella*, of which the grains of pollen, taken from antheræ full grown, but before bursting, were filled with particles of unusually large size, varying from nearly 1/4000th to about 1/5000 of an inch in length.

. . . "I observed many of them very evidently in motion: their motion consisting not only of a change of place in the fluid, manifested by alterations in their relative positions, but also not infrequently of a change of form in the particle. . . . In a few instances the particle was seem to turn on its longer axis. These motions were such as to satisfy me, after frequently repeated observation, that they arose neither from currents in the fluid, nor from its gradual evaporation, but belonged to the particle itself. . . . I remark here also, partly as a caution to those who may hereafter engage in the same enquiry, that the dust and soot deposited on all bodies in such quantity, especially in London, is entirely composed of these (active) molecules.

(*A brief Account of Microscopical observations made in the months of June, July and August* 1827, *on the Particles contained in the Pollen of Plants; and on the general Existence of active Molecules in Organic and Inorganic Bodies.* Edinburgh *New Philosophical Journal.* Vol. 5. 1828.)

P. DELSAUX EXPLAINS THE BROWNIAN MOVEMENT

In a note by Father Delsaux, for example one may read:—
"The agitation of small corpuscles in suspension in liquids truly constitutes a general phenomenon . . . it is henceforth natural to ascribe a phenomenon having this universality to some general property of matter . . . in the train of ideas, the internal movements of translation which constitute the calorific state of gases, vapours and liquids, can very well account for the facts established by experiment."

(From notes which appeared between 1877 and 1880, cited in Perrin's work, p. 309.)

GOUY EXPLAINS THE BROWNIAN MOVEMENT

I do not believe that after careful observation anyone can doubt that in the Brownian movement we are dealing, not with accidental effects due to currents, to vibrations or to differences of temperature, but rather with a normal phenomenon, produced at constant temperature, and due to the constitution of liquids. In fact the phenomenon appears absolutely and entirely regular, it is always manifested as long as the particles remain in suspension and it persists indefinitely when they are so small that they do not deposit. From another point of view, the existence of the same movement for the particles, whether gaseous liquid or solid, shows evidently that these little spheres or these particles do not play an essential part in the movement, but only make manifest the internal agitation of the liquid. The Brownian movement therefore shows, not indeed the movement of molecules, but something which comes very near to it, and furnishes us with a direct and visible proof of the real exactness of our hypotheses concerning the nature of heat.

(Note sur le Mouvement Brownien. Par M. Gouy. *Journal de Physique*. 1888. p. 561.)

OSTWALD'S VIEWS ON THE 'REALITY' OF ATOMS

The results at which he (Berzelius) and all successors arrived was ever the same; they found that the requirements of the atomic theory are always strictly fulfilled.

Such a broad and far-reaching agreement of the empirically determined abstract laws with the atomic hypothesis formed to explain the cause of these laws justifies us in expecting still further concordance between this hypothesis and experience. In fact all our chemical experience harmonises with the atomic theory and finds in it an easy mode of expression, so that in what follows we shall always employ it. But once for all be it understood that the atomic hypothesis is only a mode of picturing to ourselves what we know of the behaviour of substances. What the "real" nature of matter is, is to us a matter of complete ignorance as it is of complete indifference.

(*Outlines of General Chemistry*, by Wilhelm Ostwald. Tr. James Walker. Macmillan. 1890. p. 5.)

THE BROWNIAN MOVEMENT HELPS TO DEMONSTRATE MOLECULAR REALITY

Comparison of all the values obtained

A table will serve to recapitulate usefully the various phenomena which enable N (the number of molecules in one gram-molecule of a gas, e.g. 2 grams of hydrogen) to be calculated, which taken altogether form what may be termed the proof of *molecular reality*.

PHENOMENA STUDIED		N.10²²
Viscosity of gases taking into account	the volume of the liquid state	45
	the dielectric power of the gas	200
	the exact law of compressibility	60
Brownian Movement	Distribution of uniform emulsion	70.5
	Mean displacement in a given time	71.5
	Mean rotation in a given time	65
Diffusion of dissolved substances		40 to 90
Mobility of ions in water		60 to 150
Brightness of the blue of the sky		30 to 150
Direct measurement of the atomic charge	Droplets condensed on the ions	60 to 90
	Ions attached to fine dust-particles	64
Emission of α-projectiles	Total charge radiated	62
	Period of change of radium	70.5
	Helium produced by radium	71
Energy of the infra-red spectrum		60 to 80

. . . "I think it impossible that a mind, free from all pre-conception, can reflect upon the extreme diversity of the phenomena which thus converge to the same result, without experiencing a very strong impression, and I think that it will henceforth be difficult to defend by rational arguments a hostile attitude to molecular hypotheses. . . ."

(*Brownian movement and Molecular Reality*, by M. Jean Perrin. Tr. F. Soddy. 1910. p. 90–91.)

MENDELÉEFF'S PERIODIC LAW

". . . It was in March 1869 that I ventured to lay before the then youthful Russian Chemical Society, the ideas upon the same subject which I had expressed in my just written *Principles of Chemistry*.

Without entering into details I shall give the conclusions I then arrived at in the very words I used:—

1. The elements, if arranged according to their atomic weights, exhibit an evident *periodicity* of properties.

2. Elements which are similar as regards their chemical properties have atomic weights which are either of nearly the same value (e.g. platinum, iridium, osmium) or which increase regularly (e.g. potassium, rubidium cæsium).

3. The arrangement of the elements, or of groups of elements in the order of their atomic weights, corresponds to their so-called *valencies* as well as, to some extent, to their distinctive chemical properties—as is apparent, among other series, in that of lithium, beryllium, barium, carbon, nitrogen, oxygen and iron.

4. The elements which are the most widely diffused have *small* atomic weights.

5. The *magnitude* of the atomic weight determines the character of the element, just as the magnitude of the molecule determines the character of a compound.

6. We must expect the discovery of many yet unknown elements—for example elements analogous to aluminium and silicon, whose atomic weight would be between 65 and 75.

7. The atomic weight of an element may sometimes be amended by a knowledge of those contiguous elements. Thus, the atomic weight of tellurium must lie between 123 and 126 and cannot be 128.

8. Certain characteristic properties of the elements can be foretold from their atomic weights. . . .

We now know three cases of elements whose existence and properties were foreseen by the instrumentality of the periodic law. I need but mention the brilliant discovery of gallium, which proved to correspond to eka-aluminium of the periodic law by Lecoq de Boisbaudran; of *scandium*, corresponding to eka-boron, by Nilson; and of germanium, which proved to correspond in all respects to eka-silicon, by Winkler.

(*The Periodic Law of the Chemical Elements*, by Professor Mendeléeff. Faraday Lecture. 1889.)

CHAPTER TWENTY

Man Learns About His Body

The knowledge of the human body

The understanding of the working of the human body is the most difficult and most important of human tasks. A precise understanding of the processes that go on in the human organism would probably indicate the way to maintain health, delay old age, cure most (if not all) diseases, and even to modify the form, stature and faculties of the human beings of generations to come. Only a part, and that a small part, of this knowledge is now ours, but it is increasing at an ever greater rate; and Man may be forced, even during the next century, to assume the responsibility for allowing, refusing, or controlling unheard-of modifications of his biological habits, on which society is based and by which its continuance has been assured.

Microscopic anatomy

The study of the human body is often divided into *Anatomy*, the description of its structure, and *Physiology*, the description of its working. Anatomy can be carried to great perfection without much knowledge of physiology, and during the eighteenth century the anatomy of the human body, as far as it is observable by eye, was very well and skilfully mapped. About the year 1800 it was recognised that the body was made up of different kinds of material (muscle, membrane, cartilage, etc.), and these were called *tissues*. Microscopes were greatly improved in the years after 1830, and so the structure of these tissues, as far as the microscopes of the period would reveal it, was gradually made clear. The great discovery that arose from this was that *all living organisms are either cells or communities of cells*, each cell being a unit of life capable of independent existence in a suitable environment. In 1838–9 the cell-theory was put forward in a simple form, both for plants and animals.

Soon after it became clear that in every cell there was an essential living stuff, a sort of slime or jelly, which was termed *protoplasm*. There was then no technique for discovering what it was. Some workers thought it was a chemical compound, others believed it to be a mixture of compounds in solution. To-day we know that the problem of the nature of protoplasm is exceedingly complicated. Each organism has a protoplasm in some way different: it is not a homogeneous substance, it contains many different chemical entities, and is a complex system in chemical and physical equilibrium. We know that we have not yet the technique to discover the working of this basic matter of life, but we do not regard as for ever insoluble the problem of explaining the behaviour of protoplasm, in so far as it can be explained in terms of physics and chemistry; for whether living matter has or has not special properties that can only be expressed in terms of life, we must suppose that it also conforms to the laws of chemistry and physics.

Cells were at first thought to be little more than tiny bladders filled with structureless protoplasm; but after the eighteen-seventies, when the microscope was brought to the stage of showing all that light can show, it became clear that they had a very complicated anatomy indeed, especially in respect of the arrangements which make each cell characteristic of the species of organism and of the individual of which it forms a part. Thus every cell in my body contains physical structures which not only characterise it as that of a man, and not of a monkey or a bull or a snake, but also characterise it as a cell the particular man Sherwood Taylor and no other*; and these structures somehow enable it to reproduce other cells of exactly that kind. We know that certain parts of the cell have certain functions in cell-division, heredity and so forth, but it is not far from the truth to say that we do not know in terms of chemistry and physics how anything in the cell works.

Cytology, the study of cells

The study of cells has made great advances in the direction of describing cells and their behaviour: when a technique

* I am not a twin.

is found for studying chemical reactions in spaces so small that the microscope can scarcely reveal them, then we shall see a gigantic increase in our knowledge of the human body and in the means of controlling it. Cancer is a cell-disease; the whole constitution of the individual derives from the germ-cells from which he is formed; what, in medicine, could be greater than the power to prevent or cure cancer, what, in statecraft, than the power to control and produce racial characters?

Physiology

The anatomists of the eighteenth century had a precise and detailed knowledge of the various organs of the body; they knew fairly well what those organs were for, but they had only the vaguest ideas of how they did their work. This is not surprising, for the body is chiefly to be explained in terms of chemical and electrical actions, neither of which was then sufficiently understood. Organic chemistry remained very unsystematic till the 'seventies and as most of the chemistry of the body is a very complex organic chemistry, not much progress was made in elucidating it until the twentieth century.

Most physiological experiments require surgical operations on animals and the technique of surgery was very crude until the work of Lister made it one of the principal departments of medicine. This is another reason why physiology made little progress until near the end of the nineteenth century.

The food we eat

The study of food was one of the simpler problems because it can be carried on without surgical intervention, but it could not go far until the chemical nature of the chief food-stuffs was known. Lavoisier understood that food underwent a sort of combustion in the body, so producing animal heat; but his experiments were interrupted by the revolutionaries who put him to death. Then, about 1840, Liebig was so bold as to break away from the idea that the ordinary operations of chemistry and physics were suspended in the body by a mysterious vital force, and attempted to investigate the chemistry of the life-process. He distinguished proteins, carbohydrates and fats as constituents of animal and plant tissues. He realised that plants made these from simple inorganic compounds and that animals obtained them from plants, but he did not understand that animals, by digestion, broke down their plant-food and built it up into *different* animal proteins, carbohydrates and fats.

The action of the digestive organs was worked out in the mid-nineteenth century. A famous piece of work was that of Beaumont who at various times between 1825 and 1833 had the good fortune to be able to study a patient who had, as a result of a gun-shot wound, a permanent opening into his stomach. This enabled Beaumont to study the gastric juice and digestion (p. 321). Gradually the functions of the other parts of the digestive system were pieced together, and before 1900 our physiological and chemical ideas about the breaking down and absorption of food-stuffs were fairly complete. But long before this, the idea of the conservation of energy had become influential in science, and the fact that animals were constantly producing heat and work, made it seem pretty certain that one of the chief functions of food-stuffs was to be oxidised to produce energy (pp. 167, 322). It was by no means easy to show how or in what parts of the body this was done. In 1847 von Helmholtz, who touched on almost every department of science and always made a significant contribution, proved by a delicate electrical thermometer that the muscles gave out heat when contracting; and other workers gradually proved that a muscle was, as it were, an engine in which glycogen (a substance built up from the sugar which digestion makes from, food carbohydrates) breaks down to lactic acid, carbon dioxide and energy, which appears partly as work, partly as heat. Oxygen from the blood stream then oxidises part of the lactic acid, liberating energy and so restoring part of the glycogen. This was worked out about 1902, but it was not necessary to know all this in order to study what food human beings required. Attempts were made to measure the chemical energy of food-stuffs (by combustion) and to make a sort of balance-sheet of energy taken in and given out by the organism. This led, between 1885 and 1905, to the study of the calorific value of food-stuffs and the calories of energy needed by workers of various kinds. Large calorimeters were made in which men could live and work, and in this way we came to know for the first time how much food and what kind of food we need. It appeared that certain prison diets, and even military diets, had been slow starvation: the "thick and nourishing gruels" prescribed for the starving poor in the hungry 'forties were seen to be useless for preserving life. The immense improvement in the nourishment of modern children is, in part at least, due to these researches.

But such experiments as these revealed a strange thing. Certain diets which were composed of carefully purified food-stuffs and which had a fully sufficient calorie value were found to be incapable of keeping animals in growth and health, but the addition to these of quite inconsiderable quantities of certain natural substances, such as milk, rendered these diets healthful, and restored the animals that had suffered from them. From about the year 1906 it became clear that in addition to proteins, carbohydrates, and fats, animals required very minute quantities of certain chemical substances which were not in the usual sense food-stuffs at all. These are called *vitamins* and the first of them were isolated in fairly pure condition about 1912. Soon it was discovered that a number of hitherto mysterious diseases—rickets, scurvy, beri-beri, pellagra, etc.—were due to the lack of these vitamins, and now that most of the vitamins have been synthesised and are easily available, these diseases need only exist where medical attention is unavailable or unsought.

The integration of the body

The body is not merely a package of organs. It acts as a whole and a number of its organs usually have to co-operate, without the conscious control of the subject, to bring about such essential activities as respiration, digestion or reproduction.

Thus, for example, violent exercise causes rise of temperature, increased breathing, more rapid action of the heart: digestion involves the successive and unconscious muscular action of the stomach and bowels and the outpourings of the secretions of numerous glands. All this was known early in the nineteenth century, but the mechanism by which these organs are co-ordinated was discovered but slowly. That the nerves, spinal cord and brain are means of communication between organs seemed obvious, for the nerves were known to be connected to every organ and had long been known to cause the muscles to contract.

During the greater part of the nineteenth century physiologists were working out the interconnection of the nerves in the 'telephone exchange' of the spinal cord and brain. From 1811 there were attempts to indicate the parts of the brain which controlled certain organs, chiefly by noticing what failures of function resulted from disease or accidental damage in localised areas of the brain. Experiments on the brains of animals were beyond the technique of the time. Gall about 1811 went beyond reasonable deduction, and supposed that the 'bumps' of the skull indicated high development of certain parts of the brain corresponding to various mental functions; he thus founded the pseudo-science of phrenology. There were, however, some real advances. Sir Charles Bell, the surgeon, in 1820 discovered the separation of the sensory and motor nerves: while Johann Müller in the 'thirties discovered the law of specific nerve energies—that stimulation of the optic nerve could give no sensation but light, that stimulation of the auditory could cause no other sensation but that of sound, and that in fact each nerve-fibre could cause one and only one sensation or movement. The law was of great importance as demonstrating that only certain modes of knowledge of the outer world are possible and that these depend on the construction of our human organs. This was very soon followed by the discovery that certain nerves regulated the rate of heart-beat and the flow of blood through the vessels, and in 1851 it was shown that nerves controlled the secretion of the salivary glands.

The brain, however, remained largely uncharted. Broca in 1861 had shown that it contained an area, damage to which prevented speech and nothing else; and in the 'seventies the mapping of areas corresponding to the different bodily functions had begun and has since greatly advanced. But no one has as yet attributed *intellectual* functions to particular areas and there is no certainty as to what part of my brain I am using when I compose these paragraphs, though we are fairly confident in indicating the part that operates the muscles of my hand as it writes.

In 1833 Marshall Hall described a *reflex*, a process by which a stimulus produces a reaction independent of conscious will, and towards the end of the century the study of reflexes as connecting together and *integrating* a series of complicated motions such as are involved in swallowing or walking, was worked out by Sir Charles Sherrington. After 1900 Pavlov showed by lengthy experiments on dogs how these reflexes could be *conditioned*, connected with certain parts. Thus the approach of food makes a dog's mouth water. If a dinner-bell accompanies the food on several occasions, the dinner-bell will soon by itself make the dog's mouth water. Elaborate studies, of which the above is a very simple example, give a knowledge of the action of the brain which is *scientific*, because based on reasoning about reproducible observations; but how far they can ever be applied to the higher mental functions we cannot guess.

A quite different approach to the problem of nervous action has been that of *psychology*, which can proceed in two chief ways, (1) the study of what we perceive as going on in our mind, (2) the study of the behaviour of men and animals. The first may be and often is a useful method, but it is not strictly scientific, because the evidence from which it starts is not verifiable or capable of being precisely recorded. The second, which proceeds by experimental observations of animal or human actions (e.g. studies of cats finding their way out of puzzle-boxes or of rats passing through mazes, or of the influence of fatigue or of alcohol on the performing of some simple task), provides data which can be numerically observed

MAN LEARNS ABOUT HIS BODY

and mathematically treated by statistics so as to yield scientific laws. This *experimental psychology* began in the eighteen-seventies and has become very influential in the last thirty years, when it has been applied, with some degree of success, to the study of human societies.

Integration through the glands

It had been known from very ancient times that certain organs had powerful effects on others remote from them, e.g. that the castration* of male animals prevents them from developing the characteristics peculiar to adult males of the species, e.g. the large comb of the cock, the beard and deep voice of a man. No one knew how this came about. Further interest in the functions of the ductless glands appeared about 1855 when Addison discovered a strange and fatal disease which accompanied the destruction of the adrenal glands. Surgical removal of these glands or of the thyroids from animals proved fatal, but surgery was then so often fatal that this result attracted little attention. After Lister's technique was established, operations were performed for the cure of goitre enlargement of the thyroid gland; and it was found that the total removal of this gland was followed by a curious disease characterised by sluggishness and imbecility, low temperature, and a peculiar thickening of the tissues. In 1888–90 this disease, *myxœdema*, was successfully treated by administering raw thyroid gland to the patients (p. 318). Clearly the thyroid, like the testes, influenced the whole body. There are parts of the world where iodine is totally lacking in the drinking water, and here the thyroid, which normally contains iodine, is unable to function properly, and the thyroids of children born of parents suffering from this condition may fail to develop at all. Such children grow up as stunted, mindless, infantile idiots called cretins: these, while still in infancy, could, it was found, be entirely cured by administering thyroid gland. About 1890 the idea was put forward that ductless glands normally produced minute quantities of substances which were carried

* Removal of the sex-glands.

to distant organs by the blood stream, and caused them to exert their functions. These substances were thought of as 'chemical messengers' and after 1900 were called *hormones*. Since that time they have been greatly studied.

The development of our knowledge of the ductless glands came about chiefly through surgical technique. As long as operations were septic and surgery rough and ready, the removal of internal organs from animals was usually quickly followed by their death: but to-day even such organs as the pituitary gland, tucked away beneath the brain, can be removed without injury to other organs, and its peculiar effects discovered. The result of such studies has been to show that a number of glands—the testis and ovary, the thyroid, parathyroid, adrenal and pituitary glands, all produce minute quantities of one or more chemical compounds which have profound influences upon other organs. These hormones influence and control metabolism, growth, puberty, the sexual cycle, pregnancy and lactation, and other fundamental bodily processes. One or two of the active principles of these hormones, notably that of the secretion of the thyroid, have been synthesised chemically. In no case do we understand how they bring about their astonishing effects, but it is clear that they are of the greatest importance in governing and controlling essential functions of the body. Thus it is pretty certain that if we had unlimited supplies of the growth-hormone of the pituitary we could make any child grow to seven or eight feet in height, and it is at least likely that by administering suitably blended hormones we could profoundly modify the temperament of individuals. No discovery yet made is more threatening to human freedom.

But setting such speculations aside there is one hormone that stands above all others in the benefit that mankind has reaped from its use: insulin, secreted by the islets of Langerhans, in the pancreas. The isolation of this by Banting in 1921 as the climax of years of experiment by himself and many others, has made it possible for many thousands of diabetics, who would formerly have succumbed after a few years, to continue indefinitely an active and, in some cases, a useful life.

Examples Concerning Physiology

THEODORE SCHWANN ON SCIENTIFIC EXPLANATIONS IN
PHYSIOLOGY

The first view of the fundamental powers of organised bodies may be called the *teleological*, the second the *physical* view. An example will show at once, how important for physiology is the solution of the question as to which is to be followed. If, for instance, we define inflammation and suppuration to be the effort of the organism to remove a foreign body that has been introduced into it; or fever to be the effort of the organism to eliminate diseased matter, and both as the result of the "autocracy of the organism", then these explanations accord with the teleological view. For since by these processes the obnoxious matter is actually removed, the process which effects them is one adapted to an end; and as the fundamental power of the organism operates in accordance with definite purposes, it may either set these processes in action primarily, or may also summon further powers of matter to its aid, always, however, remaining itself the "primum movens". On the other hand, according to the physical view, this is just as little an explanation, as it would be to say, that the motion of the earth around the sun is an effort of the fundamental power of the planetary system to produce a change of seasons on the planets, or to say, that ebb and flood are the reaction of the organism of the earth upon the moon.

In physics all those explanations which were suggested by a teleological view of nature, as "horror vacui", and the like, have long been discarded. But in animated nature, adaptation —individual adaptation—to a purpose is so prominently marked, that it is difficult to reject all teleological explanations. Meanwhile it must be remembered that those explanations, which explain at once all and nothing, can be but the last resources, when no other view can possibly be adopted; and there is no such necessity for admitting the teleological view in the case of organised bodies. The adaptation to a purpose which is characteristic of organised bodies differs only in degree from that which is apparent also in the inorganic part of nature; and the explanation that organised bodies are developed, like all the phenomena of inorganic nature, by the operation of blind laws framed with the matter, cannot be rejected as impossible. Reason certainly requires some ground for such adaptation, but for her it is sufficient to assume that matter with the powers inherent in it owes its existence to a rational Being. Once established and preserved in their integrity, these powers may, in accordance with their immutable laws of blind necessity, very well produce combinations, which manifest, even in a high degree, individual adaptation to a purpose.

(*Microscopical Researches into the accordance of the Structure and Growth of Animals and Plants*. Theodore Schwann, originally published 1839, translated by H. Smith, 1847.)

WILLIAM BEAUMONT EXPERIMENTS ON THE LIVING STOMACH

[Alexis St. Martin was accidentally wounded by the discharge of a musket, on the 6th of June, 1822. The charge consisting of powder and buckshot, was received in his left side by the youth, he being at a distance of not more than one yard from the muzzle of the gun. Among other serious injuries his stomach was perforated, and even after he recovered perfect health, there remained an opening, normally closed by a valve-like formation of the tissues, but through which the stomach could be inspected, and matters introduced and withdrawn. William Beaumont, whose surgical treatment had saved St. Martin's life, took him into his service, and in 1825 and 1832-3 carried out extensive researches upon digestion in general and the relative digestibility of food-stuffs.]

EXPERIMENT 40

March 18th. At 9 o'clock a.m. he breakfasted on *soused tripe* and *pig's feet*, *bread* and *coffee*.

At 9 o'clock 30 minutes, took out and examined a portion,— found it in a half-digested condition, tripe, pig's feet, and bread, all reduced to a pulp, floating in a large portion of fluids. Placed it on the bath.*

At 10 o'clock, examined stomach again,—tried to extract another portion,—could find little or no chyme,—a very little gastric juice, with a few small fibrous particles of tripe, and some coffee-grounds. His breakfast appeared to have been digested and had passed from his stomach in *one hour*.

The portion first taken out and placed on the bath was also, at the end of one hour, reduced to nearly a complete chymous condition, a very few of the small particles of tripe and coffee-grounds only left, as in the stomach.

EXPERIMENT 65

To ascertain whether the sense of hunger would be allayed without the food being passed through the œsophagus,† he fasted from breakfast time till 4 o'clock p.m. and became quite hungry. I then put in at the aperture three and a half drachms of *lean boiled beef*. The sense of hunger immediately subsided, and stopped the borborygmus or croaking noise caused by the motion of air in the stomach and intestines, peculiar to him since the wound, and almost always observed when the stomach is empty. . . .

(*Experiments and Observations on the Gastric Juice and the Physiology of Digestion*. William Beaumont, M.D. Edinburgh, 1838.)

ROBERT MAYER IN 1842 APPLIES THE CONSERVATION OF ENERGY TO ANIMALS AND PLANTS

The second question refers to the cause of the chemical tension‡ produced in the plant. This tension is a physical§ force. It is equivalent to the heat obtained from the combustion of the plant. Does this force, then, come from the vital

---

* Water-bath.

† Gullet.

‡ We would say 'chemical energy,' but this idea hardly existed in Mayer's day.

§ i.e. a force of nature—not a physical force as opposed to a chemical force.

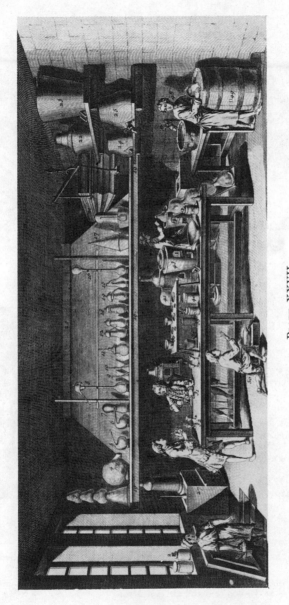

PLATE XXVII

Laboratory of early eighteenth century.

1

2

3

4

PLATE XXVIII

Photographs of the myxœdema patient whose treatment is described on page 324.

(These 4 illustrations are reproduced by courtesy of Dr. Hugh M. Raven and the *British Medical Journal*).

processes, and without the expenditure of some other form of force? The creation of a physical force, of itself hardly thinkable, seems all the more paradoxical when we consider that it is only by the help of the sun's rays that the plants perform their work. By the assumption of such a hypothetical action of the "vital force", all further investigation is cut off, and the application of the methods of exact science to the phenomena of vitality is rendered impossible. Those who hold a notion so opposed to the spirit of science would be thereby carried into the chaos of unbridled phantasy. I therefore hope that I may reckon on the reader's assent when I state as an axiomatic truth, *that during vital processes a conversion only of matter, as well as of force, occurs, and that creation of either the one or the other never takes place.*

The physical force collected by plants becomes the property of another class of creatures—of animals. The living animal consumes combustible substances belonging to the vegetable world and causes them to reunite with the oxygen of the atmosphere. Parallel to this process runs the work done by animals. . . . In the animal body chemical forces are perpetually expended. Ternary and quaternary* compounds undergo, during the life of the animal, the most important changes, and are, for the most part given off in the form of binary compounds,† as burnt substances. The magnitude of these forces, with reference to the heat developed in these processes, is by no means determined with sufficient accuracy,‡ but here where our object is simply the establishment of a principle it will be sufficient to take into account the heat of combustion of the pure carbon. . . . (*Calculations follow*). . . . If the animal organism applied the disposable, combustible material solely to the performance of work the quantities or carbon just calculated would suffice for the times mentioned. In reality, however, besides the production of mechanical effects there is in the animal body a continuous generation of heat. The chemical force contained in the food and inspired oxygen is therefore the source of *two* other forms of power, namely

---

* i.e. complex organic compounds, containing three or four elements.

† e.g. carbon dioxide and water.

‡ Thermochemistry was scarcely begun.

mechanical motion and heat; and the sum of these physical forces produced by an animal is the equivalent of the contemporaneous chemical process. Let the quantity of mechanical work performed by an animal in a given time be collected and converted by friction or some other means into heat; add to this the heat generated immediately in the animal body at the same time, we have then the exact quantity of heat corresponding to the chemical processes that have taken place.

In the active animal the chemical changes are much greater than in the resting one. Let the amount of the chemical processes accomplished in a certain time in the resting animal be x, and in the active one be $x + y$. If during activity the same quantity of heat were generated as during rest, the additional chemical force y would correspond to the work performed. In general, however, more heat is produced in the active organism than in the resting one. During work, therefore, we shall have x + a portion of y heat, the residue of y being converted into mechanical effect. The maximum mechanical effect produced by a working mammal hardly amounts to one fifth of the force derivable from the total quantity of carbon consumed. The remaining four-fifths are devoted to the generation of heat.

(*Organic Motion and Nutrition*. Robert Mayer. Originally published in 1842, quoted from J. C. McKendrick's *Hermann von Helmholtz*. Fisher Unwin. 1899.)

GLAND-THERAPY FOR MYXŒDEMA
(see Plate XXVIII)

Mrs. S. was born in May 1829, therefore when she died in January 1924 she was over 94 years of age. The symptoms of myxœdema began in about 1870, and were untreated for twenty years, by which time she was bedridden and imbecile, and her appearance is to be seen in Photograph 1. After five weeks treatment with thyroid tablets her mental condition was much improved and the condition of the skin generally was clearing up (Photograph 2). After fifteen months treatment she was practically normal again (Photograph 3): the regrowth of her hair is to be noted. Within a year or two she was well enough to nurse her daughter through an attack of typhoid and lived to a ripe old age—happy, healthy and mentally active (Photograph 4).

The Life-History of a Case of Myxœdema. Hugh M. Raven. *British Medical Journal*, Oct. 4th, 1924, p. 622.)

CHAPTER TWENTY-ONE

Waves and Particles

Eighteenth century views of Light

The two views of light held in the eighteenth century were the corpuscular theory, which treated light as a stream of particles, and the wave theory, which treated it as undulations in some medium. Newton did not decide definitely for or against either view. He supposed that waves in the ether existed, but supposed light to consist of particles capable of setting up such waves. His criticisms of the wave-theory were directed at the form of it originated by Huyghens, who supposed light to consist of *longitudinal* waves in the æther, like those of sound in air.

That light had a definite velocity and was not transmitted instantaneously had appeared from the work of Römer. The times of the eclipsing of Jupiters satellites by their parent planet could be pretty exactly calculated, yet it appeared that these eclipses were twenty-two* minutes later when Jupiter was furthest from the earth than when it was at its nearest. The period of twenty-two minutes was interpreted as the time required for light to cross the earth's orbit. Before 1727, James Bradley noticed a very important phenomenon. The stars appeared to show a very small regular yearly displacement, back and forth. This could be accounted for (p. 332) by supposing that the apparent direction of the light from a star at any time was a combination of the real velocity and direction of the light and the velocity and direction of the earth. This phenomenon Bradley called the aberration of light.

Bradley followed the corpuscular theory, and so this explanation presented no difficulties, so long as it was assumed that the particles of light were not attracted by the earth. But those who believed that light consisted of waves in a medium could

---

* The correct figure is sixteen minutes.

only explain aberration if they supposed the medium to be quite undisturbed by the passage of the earth through it,—a very important conclusion.

The weight of eighteenth-century opinion was in favour of the corpuscular view and light was treated as an imponderable chemical element, but in 1801 Thomas Young revived the undulatory theory in order to explain the interference of light—the fact that two beams of light could combine to form darkness. Two waves could neutralise each other, but two particles could not. Young thought, at first, that light consisted of longitudinal waves like those of sound, and so could not explain some phenomena, such as polarisation. But in 1817 he considered the possibility of transverse waves, by which all the well-known optical effects were readily explained and new ones predicted. Thus William Rowan Hamilton calculated from this theory that under certain circumstances a ray of light could be refracted into a cone of light. This surprising prediction was verified by experiment and seemed to confirm the theory very strongly.

We have seen (pp. 256–57) how it was established by Clerk Maxwell and others that light was not a mechanical vibration, but a rapid alternation of electrical and magnetic fields. When the waves were thought of as mechanical transverse vibrations, there obviously had to be an æther to vibrate: but it was not so obvious that electrical and magnetic fields required a material substratum. Nevertheless, this was the usual mode of thought up to about 1910.

Theories of the ether

Huyghens's idea of the ether was like the first matter of Descartes,—a mass of minute elastic spheres, much smaller than the atoms of matter: the longitudinal waves of his theory were transmitted by the collisions of these spheres. But Young's transverse waves could only be transmitted by an elastic solid medium, so from about 1830 the ether was thought of as an elastic solid resembling a jelly. Yet Bradley's observations had showed that the ether was not disturbed by the earth's motion, moreover the motion of the earth and planets was evidently not impeded by it. A remote analogy to this was found in semi-solids like pitch, which can vibrate like a solid but yet can flow like a liquid if given sufficient time. Thus ether might vibrate like a solid at the high frequencies of light-vibrations, but yet allow the planets to slip through it. The physicists of the nineteenth century wanted to figure out a *structure* for the ether that would allow of this, and a great deal of work was done in the effort (which had some success) to work out a mechanical structure for æther that would allow it to have a rotational elasticity but no resistance to non-rotational movement, so allowing light to be transmitted, but allowing the planets to move freely.

All these theories required that the ether should be made up of particles of some kind and this raised an insoluble difficulty. The particles of ether must, it seemed, interact with those of matter. But the development of the kinetic theory indicated that the necessarily very fine particles of ether ought to absorb almost the whole of the energy from matter. Thus the notion of the ether presented great difficulties. It was hoped that these might gradually be solved, but they increased rather than diminished.

The Michelson and Morley experiment

In the eighteen-eighties the most serious difficulty concerning the ether was posed by the experiments of Michelson and Morley (pp. 333–34). The velocity of light was measured along the path of motion of the earth and at right angles thereto. Now the light was supposed to be a wave-motion in the ether, and Bradley's observations showed that the ether did not move with the earth. Thus the ether ought to be moving relatively to the earth at about twelve miles a second. Now, the velocity of light along the direction of the earth's motion should on this basis be different from its velocity across the earth's motion, just as a swimmer takes longer to swim first a mile up and then a mile down a river, than to swim two miles across it. But no difference was found. Michelson and Morley's experiment seemed to prove that the ether was stationary relative to the earth, or, in other words, that it moved with it, yet Bradley's observations on the aberration of light proved that it did not move with the earth. The difficulty was temporarily resolved by the assumption of the 'Fitzgerald contraction': it was supposed that the ether was stationary and that when bodies are in motion relative to the ether they contract slightly, just enough to neutralise the effect that the Michelson and Morley experiment was expected to show and did not. But no one could find any corroborative evidence of this hypothesis of contraction, and the solution did not seem a very satisfying one.

So even in the eighteen-nineties the ether seemed a very puzzling and elusive conception, but up to 1900–1910 there were no serious doubts that light could be completely explained as a transverse wave-motion in the ether. Obviously this theory gave a close approximation to most of the facts, but the part of it that was effective was the wave-aspect and not the idea of ether.

Relativity Theory

In 1905 Einstein propounded his theory of relativity (p. 335) and showed that the difficulties raised by the Michelson-Morley experiment could be met by a new statement of the meaning of space and time. Later he treated the world as a four dimensional continuum, one of the dimensions being time, and was able to include optical, electric and gravitational phenomena in one statement. He showed that all the phenomena of nature went on just as if there were no ether at all, and that if it were assumed that 'space could transmit electromagnetic waves', the ether was an unnecessary assumption. Einstein's theory was, of course, not a mere re-interpretation of known facts: it led to a number of important predictions which could be confirmed experimentally e.g. that light was deflected in a powerful gravitational field, that spectral lines were shifted by a strong gravitational field, that the mass of a body was increased by an increase in its velocity, that mass and energy are interconvertible,—all of which were confirmed by experiment. Relativity theory is now an essential part of the scientific world-view.

Quantum theory

From the eighteen-nineties it became apparent that there were many phenomena that could not be explained in terms of the classical electromagnetic theory of radiation. These were concerned with very small-scale phenomena e.g. the emission and absorption of light by atoms.

Thus the 'classical' theory of electro-magnetic waves could not explain the fact that the spectra of glowing gases consisted of light of a number of fixed wavelengths only; nor could it show why the radiation from a hot black body contains a very small proportion of the very short wavelengths, whereas on the classical theory these should be present in the greatest proportion. These and other problems were solved by Planck's quantum-theory, put forward in 1900, which stated that a vibrating system (oscillator) of given frequency could not have any given energy, but only the values represented by $nh\nu$ where n is a whole number, h a very minute constant (6.55×10^{-27} erg-seconds) and ν is the frequency of oscillation. In the case of an oscillator with a very minute energy, such as an atom or electron, this meant that it could only gain or lose energy by relatively large jumps. Thus it might have energy nil, $h\nu$, $2\ h\nu$, $3\ h\nu$, but no intermediate values. The amount of energy represented by $h\nu$ was called a quantum: so when the energy of the oscillator changed it absorbed or gave out one or more whole quanta. Einstein added to Planck's notion the idea that these quanta were not simply quantities of energy that could diffuse out into space in all directions, but that each quantum was directed in a definite direction and kept together as a sort of corpuscle or 'photon'. Thus the new conception combined the two traditional views: light and other radiations were to be regarded both as corpuscles and waves. The quantum theory became very important from 1913 when Niels Bohr (p. 301) made it the basis of his theory of the atom. In 1924 Louis de Broglie made a great advance by treating the ultimate constituents of *matter*, e.g. electrons, as at once corpuscles and waves; and in 1927 Davisson and Germer, and later other workers, showed experimentally that electrons had some of the properties of waves, e.g. that they exhibited diffraction phenomena.

From this period De Broglie, Schrödinger, Heisenberg, Dirac, and others developed the method of wave-mechanics or quantum-mechanics, which seems to be the most fundamental way of treating natural phenomena. Both radiation and matter are treated as wave-packets and enormous successes have resulted in all matters concerning the interaction between them. For small-scale phenomena, such as the interactions of atoms in the molecule, it is the only mode of predicting correct results. Waves, however, cannot indicate an exact position, so wave-mechanics cannot tell us the precise position of a particle but only the varying probability of its being in any specified place. But no other method is available for studying these phenomena, and so we find that we have to accept the 'uncertainty principle',—that *we cannot observe or theoretically predict the precise positions and velocities of particles.* Thus we have had to abandon the traditional form of the atomic theory, in which the world was supposed to consist of particles with precise shapes, sizes, positions and motions, and our modern world-view is of space-time wherein there is a continuously varying probability which manifests as the particles of which we regard the world as being constituted. Wave-mechanics has been extremely fruitful in all calculations concerned with small-scale phenomena, but it does not apply only to these phenomena; for, when we deal with very large assemblages of 'particles', as in ordinary physics, the wave-mechanical laws come to coincide with the classical laws. We shall doubtless see great developments in this field, and the present wave-mechanics is not the last word. It works admirably, but there is no agreement about the physical meaning of its symbols. But the fundamental idea that the world is not composed of two quite different entities, waves of radiation and particles of matter, will undoubtedly remain as one of the great simplifications of science.

The unification of the world of science

Contrast for a moment the views of 1890 with those of 1948. In 1890 we had seventy or eighty quite different kinds of atom, we had ether, electricity, radiation, gravitation, magnetism, chemical affinity—all different and not derivable from a common source. To-day we have four-dimensional space-time whose curvature expresses all our notions of force, we have the probability-waves of quantum-mechanics giving rise to the 'particles' and 'waves' from which all atoms and all radiation are derived. There remains the problem of combining the point of view of relativity and that of quantum theory, towards which something was done by the late A. S. Eddington. This process might give us a view of all phenomena as manifestations of one entity, which is the final goal of theoretical science.

Examples on Waves and Particles

JAMES BRADLEY EXPLAINS ABERRATION

. . . "I considered this Matter in the following Manner. I imagined CA to be a Ray of Light, falling perpendicularly

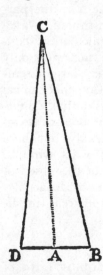

upon the line BD; then if the Eye is at rest at A, the object must appear in the direction AC, whether Light be propagated in Time or in an Instant. But if the Eye is moving from B towards A, and light is propagated in Time, with a velocity of the Eye as CA to BA; then Light moving ırom C to A, whilst the Eye moves from B to A, that particle of it by which the object will be discerned, when the Eye in its Motion comes to H, is at C when the Eye is at B. Joining the points B, C, I supposed the line CB, to be a Tube (inclined to the line BD in the Angle DBC) of such a Diameter as to admit of but one particle of Light; then it was easy to conceive that the Particle of Light at C (by which the object must be seen when the Eye, as it moves along arrives at A) would pass through the tube BC, if it is inclined to BD in the Angle DBC, and accompanies the

Fig. 52.—Bradley's illustration of the Aberration of Light.

Eye in its Motion from B to A; and that it could not come to the Eye placed behind such a Tube, if it had any other Inclination to the Line BD . . . (*hence he shows that*) the Sine of the Difference between the real and visible Place of the Object, will be to the Sine of the visible Inclination of the Object to the Line in which the eye is moving, as the velocity of the eye to the Velocity of Light.

(Philosophical Transactions. 1727–8. No. 406. *A letter from the Reverend Mr. James Bradley, Savilian Professor of Astronomy at Oxford, and F. R. S. to Dr. Edmond Halley Astronom. Reg. &c., giving Account of a new discovered motion of the Fix'd Stars.*)

THE MICHELSON-MORLEY EXPERIMENT

. . . "The discovery of the aberration of light was soon followed by an explanation according to the emission theory. The effect was attributed to a simple composition of the velocity of light with the velocity of the earth in its orbit. The difficulties in this apparently sufficient explanation were overlooked until after an explanation on the undulatory theory of light was proposed. This new explanation was at first almost as simple as the former. But it failed to account for the fact proved by experiment that the aberration was unchanged when observations were made with a telescope filled with water. For if the tangent of the angle of aberration is the ratio of the velocity of the earth to the velocity of light, then, since the latter velocity in water is three-fourths its velocity in a vacuum, the aberration observed with a water telescope should be four-thirds of its true value.

On the undulatory theory according to Fresnel, first; the ether is supposed to be at rest, except in the interior of transparent media, in which, secondly, it is supposed to move with a velocity less than the velocity of the medium in the ratio $\frac{n^2 - 1}{n^2}$, where n is the index of refraction.

(There follows the account of the experiment (p. 328)

. . . the displacement to be expected was 0.4 fringe.* The actual displacement was certainly less than the twentieth part of this and probably less than the fortieth part. . . . It appears

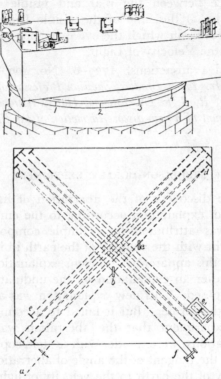

FIG. 53.—*Above:* Michelson and Morley's Apparatus. *Below:* Path of the light.

from all that proceeds reasonably certain that if there be any relative motion between the earth and the luminiferous æther it must be small; quite small enough entirely to refute Fresnel's explanation of aberration.

(*On the Relative Motion of the Earth and the Luminiferous æther*, by Albert A. Michelson and Edward W. Morley. Phil. Mag. 1887. 5 ser. Vol. XXIV. p. 449.)

* The width of the interference fringes, the displacement of which would have revealed any difference between the time the light took to two paths, along and across the earth's path, respectively.

EINSTEIN DECLARES THE IDEA OF ÆTHER SUPERFLUOUS

"We will raise this conjecture (the content of which will be hereinafter called 'The Principle of Relativity') to the rank of a hypothesis and introduce in addition a hypothesis which is only in appearance incompatible with the former hypothesis, namely, that light, in a vacuum, is always transmitted with a certain velocity V which is independent of the state of motion of the emitting body. These two premises are sufficient to establish a simple electrodynamic, free of contradictions and based on the Maxwell theory for stationary bodies. The introduction of an 'ether of light' will prove superfluous, because, according to the ideas which will here be developed, neither will there be introduced an 'absolutely stationary space' endowed with special qualities, nor will a velocity vector be co-ordinated to any point of the vacuum, in which the electromagnetic processes take place."

(*On the Electrodynamics of bodies in motion.* By A. Einstein. Annalen der Physik. 1905.17.891.)

Next year he proved from his theory that:—

"The mass of a body is a measure of the energy it contains; if the energy is varied by L, the mass changes correspondingly by $1/9.10^{20}$, if the energy is measured in ergs and the mass in grams.

"It is not impossible that the theory will be successfully tested in the case of bodies whose inherent energy can be altered to a great extent (e.g. the salts of radium)."

(*Is the inertia of a body independent of its energy?* By A. Einstein, Annalen der Physik. 1905.18.639.)

THE WAVE-MECHANICAL VIEW

. . . "Finally, there is a fourth point of view developed by Heisenberg and Bohr which is most favoured at present. . . . According to this view the wave does not represent a physical phenomenon taking place in a region of space; rather it is simply a symbolic representation of what we know about the particle. An experiment or observation never permits us to say exactly that this particle occupies this position in space and that it has this particular velocity. All that experiment can show us is that the position and velocity of the particle lie within certain limits, or, in other words, that there is a certain probability that the particle lies in a particular position and some other probability that it has a particular velocity . . . (p. 7–8).

This leads to the consequence, already foreseen by Born, that we can no longer assert that there is a rigorous determinism in Nature, for all the determinism of the old dynamics rested on the possibility of determining simultaneously the initial position and velocity of a particle,* which is impossible if Heisenberg's view is admitted. Consequently there are no longer any rigorous laws, but only laws of probability.

This method of interpreting wave-mechanics introduces many surprises. In the first place, the particles have existence, and we admit that in speaking of their number we are giving expression to something which has a definite meaning. But with Bohr's ideas it is no longer possible to hold the clear and classic picture which portrays them as very small objects having position in space, a velocity and a trajectory. In the second place, the other party in this dualism, the wave, is no more than a purely symbolic and analytic representation of certain probabilities and no longer constitutes a physical phenomenon in the old meaning of the term (p. 9).

To sum up, the physical interpretation of the new mechanics remains an extremely difficult question. Nevertheless one great fact is now well established; this is that for matter and for radiation the dualism of waves and particles must be admitted, and that the distribution of the particles in space can only be foreseen by the consideration of waves . . . (p. 10).

(*An Introduction to the Study of Wave Mechanics.* Louis de Broglie. Tr. H. T. Flint. Methuen. 1930.)

---

* Consider the extract from Laplace (p. 356) in this light.

CHAPTER TWENTY-TWO

What is Science?

The objects of Science

It may seem strange that we wait till near the close of our book to define our terms. Yet how can the reader understand what science is and does, until it has been pictured to him? Moreover scientists themselves are not fully agreed as to the purpose of science, though they know well enough how to conduct and create it. It is quite clear that Science has two main objects—to enable men to *do*, and to *know*. The first was perhaps the earliest, for it would seem that the development of crafts may be traced to an earlier period than, for example, can astronomical speculations; but in the modern experimental science of the years since 1660, both aspects have always been present and complementary. Thus we find that:

(1) The desire to know, purely for the sake of knowledge, is a motive of the man of science: the knowledge so gained may or may not be adapted to the satisfying of material needs.

(2) The wish to satisfy material needs may lead to researches which in turn contribute to pure knowledge.

It is the author's impression that the first of these motives has been far more influential than the second, though writers upon science have often held the contrary. The extract from one of the most influential writers on the nature of science, Ernst Mach, which is quoted on p. 345, puts a very usual point of view concerning the functions of science and this should be read as a preliminary; but, as we shall see, there are very numerous difficulties and problems that this passage scarcely touches upon. Mach sums up elsewhere by saying that *Science may be regarded as a minimal problem consisting of the completest possible presentment of facts with the least possible expenditure of thought.* This has been criticised by saying that it is absurd to suppose that one engages in science in order not to think, but Mach's meaning is rather that thought is to be used economically, i.e. in such a way as to make a given expenditure of it go as far as possible and have the maximum effect. None the less, the majority of scientists will not agree that this economy of thought is all that they are seeking, nor do they feel that a mere statement of the common elements in a series of phenomena is enough to explain them. The kind of explanation that most of them require must not only provide a way of classifying phenomena and reasoning about them, but must show how they work.

What then does science actually do? It collects *materials*, and from them it establishes *laws, theories* and *explanations*. Let us now look at each stage of this process.

The Materials of Science

The whole of the knowledge about what we call 'things', which knowledge is the material of science, comes from sense-impressions. Every single observation or experiment was something seen with the eye, heard with the ear, felt, smelled or tasted by someone at some time.

As we have already seen (p. 316) there are a limited number of avenues of sense. We see a thing (e.g. an orange) through a light-signal and a nerve-impulse which in some totally mysterious way is manifest as the sensation we then experience; but we have no immediate assurance that there is a real orange at all, or that if there is that it is truly represented by our image of it. We are quite certain that the source of our 'image-impression' is not wholly represented by us by our unaided senses, for we do not appreciate the ultra-violet, and infra-red rays that it emits, nor can we perceive its internal structure, nor whether the seeds in it are dead or alive—the most important thing about it. There may, moreover, be an unlimited number of properties of the object, of which our senses, aided or unaided, do not tell us and possibly could not tell us anything. We are bound to conclude indeed that we do not know what is the relation between our perceptions of the orange and the 'real' orange, the thing-in-itself, if indeed such a thing exists.

One thing is, however, quite clear: namely, that *science is in our minds*—that it is a mental study of mental impressions.

Now the most certain of these impressions is that there is a large body of persons like ourselves studying the same science and reaching the same conclusions, and although we cannot prove it, we adopt the hypothesis that these are real people like ourselves and that they agree concerning the facts of nature and agree in finding order in nature—that is to say in the sense-impressions they receive. This order is expressed in scientific laws.

This order in nature can only be perceived if observations are made according to the same rules, though not necessarily by the same technique. Every effort must be made to eliminate the personal factors of the observer. People may disagree on colours, tastes, smells, and on any elaborate judgment made by a combination of the senses (e.g. on 'beauty' or 'sanity' or 'intelligence') but they do not disagree in deciding whether or not the stationary pointer of a dial is between the graduation marked '1' and the graduation marked '2'. Science therefore draws its evidence from the matters which are easiest to observe, and where possible from pointer-readings. This eliminates the observer but does not, of course, eliminate the experimentalist, for a class of physics-students performing the same experiment may read their instruments correctly yet obtain different results. Experimental error is as far as possible eliminated by repetition of the work under varying conditions and by reporting it in such a way that it can be repeated by others, but it can never be wholly got rid of.

It is not to be forgotten at any stage that all quantities used in science are real observations made by the senses, with or without instruments, and that all descriptions must be reducible to these observations. Thus Newton said "Absolute, true and mathematical time, of itself, and from its own nature, flows equably and without regard to anything external, and by another name is called duration: relative, apparent and common time, is some sensible and external (whether accurate or unequable) measure of duration by the means of motion, which is commonly used instead of true time; such as an hour, a day, a month, a year." But in fact absolute time of this kind is unobservable, and the reckoning of time performed in scientific experiments is made by human observers using clocks and signals by which these observers can communicate with each other. By taking this fact into account Einstein arrived at his relativity theory of space and time, and this was found to agree with the observed facts more closely than did Newton's theory of space and time and it is therefore considered to be more nearly true.

It is very important to remember that the observations of science are *selected* from our sense-impressions. Science is quite unable to study even as much of any problem as we can know (see extract, pp. 346–47). Furthermore our sense-impressions are probably poor, and possibly, though not necessarily, misleading guides to the nature of the real world of things-in-themselves, if such a world exists. Pondering on their partial and imperfect character, we must begin to wonder whether the truths of science are really in the world of things-in-themselves or whether we put them into science by the way we make our observations. Most scientists would say that we do not, but Eddington, whose opinion must be respected, seemed to claim that we do. And as it is exceedingly important to know whether something that many men use as a guide for their lives is, in fact, true—i.e. gives answers which correspond to the events of a real world—it is well before proceeding to the next paragraph to read the passage on pp. 348–49, quoted from Eddington's *Philosophy of Physical Science*.

Scientific Laws

Leaving aside the question of what our observations really represent—if they represent anything beside themselves—it is certain that they are related to each other in a mathematical way. Thus when we observe the times of swings of pendulums of different lengths, we discover the time of swing (t) of a pendulum is related to its length (l) so that l always has the same ratio to t^2, all other conditions being the same and all disturbing factors (e.g. magnetic fields) being supposed absent. We can express this as a law, *that the square of the period of oscillation of a pendulum is proportional to its length*, which we can express as $t^2 = kl$. We shall of course have to define time, length, period, oscillation, pendulum, etc.—but when we have done all this we shall have a true expression of a regularity, order, or structure in our observations. And there are scarcely any observations that we can make that do not exhibit some such regularity. Where does that regularity and order come from? Is the world of things-in-themselves an orderly world arranged according to the laws of mathematics? The Greeks thought so. Plato is reported to have said, 'God ever geometrises': Kepler was certain that if he went on trying he would find regularities in the periods and distances of the planets (pp. 114–16): J. H. Jeans to-day has said that 'God is a mathematician'.

Science does not compel this conclusion, for it tells us nothing more than that there are regularities in our observations. The explanation of this is a judgment which the whole man has to make and which cannot be made by the aid of science alone.

Some of us, like Eddington, would consider that it is to think too crudely of God or too highly of mathematics to suppose that He thinks in terms of our arithmetics and geometries; others who endeavour to guide themselves solely by scientific conclusions would see no reason to bring God into the question. The Christian here says that the problem is enlightened, if not solved, by revelation. He would say that God intends to be partly known through the world He has created (Rom. i. 20) that He has therefore created a world which can reveal its Creator to the human intellect that He has likewise created, and that in fact the order is in the world and the human intellect has the means of apprehending it: but this conclusion is not a scientific one, but philosophical or theological.

The alternative theory is that there is no reason to conclude regularity or order either in the world of things-in-themselves, nor in our crude sense-perceptions, but that we create this order by our mathematical treatment; that we artificially divide up the universe (or our sensations) into regular and similar parts and then study the relationships between these artificial divisions; that we make the parts of the universe orderly and intelligible by selection, as we make the chance distribution of tea-leaves in a cup into representations of a letter or a wedding.

Laws, theories and explanations

A scientific law is a brief statement or mathematical formula which tells the *relationship* that has always been found to exist between a number of observed quantities of a specified kind. Thus the Law of Gravitation tells us that in every case studied the force of attraction between two bodies has been found or inferred to be proportional to the product of their masses and inversely proportional to the square of the distance between them.

$$F = \frac{c \cdot mm'}{d^2}$$

where F is the force, *c* the gravitational constant, *m* and *m'* the masses* and *d* the distance between them.

The men of science does not say that every body *must* obey the law of gravitation, but only that this law† accurately describes every case that has so far been investigated. Any case of a phenomenon which appeared to disagree with this law would have to be very strictly examined before we should believe that it was correctly observed, but if such a case were found and substantiated, the Law of Gravitation would then have to be modified so as to include it. Laws *de*scribe and do not *pre*scribe: yet they are not mere descriptions, or inventories of the universe. The very existence of simple laws indicates a *regularity and order in our observations*, and this, no matter how we interpret its presence there, is the most important thing in the knowledge we obtain by sense.

A *theory* is more than a law. Thus the three Laws of Constant Composition, of Multiple and Reciprocal Proportions are statements of the regularities that have been observed in the quantities of chemical elements that combine to form compounds, but Dalton's Atomic Theory is a supposition about chemical elements which implies all these laws and certain others also. It is also an *explanation*, for when we picture to ourselves real indivisible particles of the different elements

---

* But do we know the masses except from the weights? And does not the relation between mass and weight involve the law of gravitation?

† Neglecting the modifications involved in the theory of relativity.

combining together in fixed numerical proportions, we feel
that we understand what chemical combination is.

A scientific theory, then, usually includes two things, a
mental picture or *model* that the scientist feels that he under-
stands, and some mathematical or logical relationship from
which calculations or deductions can be made.

Thus the kinetic theory of gases presents us with the picture
of a gas as consisting of countless perfectly elastic particles
(molecules) at a distance from each other which is large compared with their diameter, moving with a very high velocity,
and undergoing an unlimited series of collisions with each other
and with the walls of the vessel. That is the model. From the
model and certain assumptions a mathematical theory can be
deduced which accounts for a great variety of facts about com-
pressibility, density, heat-expansion, diffusion and many other
properties of gases. The agreement of theory and observation
is not quite exact because the model is not true to nature, for
gas-molecules are not simple, spherical, elastic particles, but
very complex structures: but the model and the resulting theory
was exact enough to be the means of building up a huge mass
of calculations from which have emerged many things of theoretical and practical importance (e.g. the theory of heat, the
theory of heat-engines, and the principles of the liquefaction of
gases, which has led to a large-scale production of fertilisers,
which are doing much towards feeding the ever more crowded
world). Its simplification of the problem of the 'real' gas was
not a defect in the model but a great merit; for it gave a means
of calculation easy and correct enough for the foundation of
masses of new knowledge and industrial practice. Such simplifications are always necessary in science, for the whole of any
real problem is always too complex for the human mind
(see p. 348).

But, of late years, we have begun to ask why we need the
model at all. Our mental picture of the model contributes
nothing of value to the final result; we use it only to discard
it, and it is the mathematical reasoning that leads to the
valuable results, and this we could have without the model.
Moreover the picture we form is absurd. I picture each mole-

cule as a little transparent sphere about as big as a mustard seed, and I think of about forty of them in a cubical glass box of about 10 cm. side, and I visualise these molecules as travelling, perhaps as fast as a fly travels, for I cannot visualise any small object as moving faster. My own imagination may be defective, but certainly no one on earth can picture ten thousand million billion molecules travelling as fast as rifle bullets. Indeed it is almost as absurd as to draw a picture of a molecule as a picture of a muchness, for it is necessarily invisible, being much smaller than a wavelength of light and having no definite surface or outline that could be seen.

Yet, for all that, I believe that my picture helps me to understand gases. I feel at home with little balls bouncing about in a box: I can imagine this, and in my mind I can watch these very imperfect images of molecules and by a sort of mental trial and error I can arrive at ideas which I can then test by the mathematical treatment. These visual images, then, that we have called mental pictures or models, are as it were rough sketches, by looking at which we can arrive at good ideas; but they do not or should not form any part of the results we record as science. This distinction was apparent in the late nineteenth century, and it occurred to the men of that time that atoms themselves were only mental constructions and that we had very little reason to suppose they were anything more than an aid to thinking about chemistry and physics. As late as 1908 Ostwald, the great German physical chemist, was scoffing at believers in the reality of atoms. But then came a flood of discoveries and measurements that were perfectly interpreted on the hypothesis that real atoms existed, and did not make sense otherwise. To-day the atom seems very nearly (but perhaps not quite) as real as a glass marble. We cannot perceive individual atoms, *i.e.* receive signals direct from them, but we can perceive their individual effects, e.g. the flash, millions of times bigger than the atom, that it makes when it is projected at high velocity on to a phosphorescent zinc sulphide screen.

But a mental model cannot be useful unless we can make it behave like that which we wish to depict. Now we can only picture the things we have seen, namely bits of matter large enough to contain thousands of billions of atoms, and we can only picture them as behaving like ordinary matter, that is to say moving at low velocities under slight fields of force. But the interior of a single atom, and still more a single electron or proton or an atomic nucleus, does not consist of ordinary matter; and there is little or no resemblance between its behaviour and that of any model we can picture. It is not a collection of millions of atoms such as our models must be based upon, so why should it behave as such? In the same way there is no reason why any model we can imagine should represent the phenomena that are observed in stars and the great spaces of the universe. So in the very small-scale world we have to adopt the quantum theory and wave-mechanics, and in the very large-scale world we have to adopt relativity-theory, in place of the ordinary mechanics, and so the phenomena that these describe cannot usefully be imagined as mechanical models. So we must cease to torture our imaginations by trying to picture the necessarily unpicturable, and in place of our usual series:

OBSERVATIONS ⟶ LAWS ⟶ MENTAL MODEL ⟶ MATHEMATICAL THEORY ⟶ PREDICTION OF NEW KNOWLEDGE

we have to substitute the sequence:

OBSERVATIONS ⟶ LAWS ⟶ MATHEMATICAL THEORY ⟶ PREDICTION OF NEW KNOWLEDGE

Unquestionably this is less satisfying and limits the powers of the human imagination to figure out new possibilities, but it is the best we can do on the difficult ground that science is beginning to explore.

Examples Concerning Modern Scientific Method

THE FUNCTION OF SCIENCE ACCORDING TO ERNST MACH (1883)

In the infinite variety of nature many ordinary events occur; while others appear uncommon, perplexing, astonishing, or even contradictory to the ordinary run of things. As long as this is the case we do not possess a well-settled and unitary conception of nature. Thence is imposed the task of everywhere seeking out in the natural phenomena those elements that are the same, and that amid all multiplicity are ever present. By this means, on the one hand, the most economical and briefest description and communication are rendered possible; and on the other, when once a person has acquired the skill of recognising these permanent elements throughout the greatest range and variety of phenomena, of seeing them in the same, this ability leads to a *comprehensive, compact, consistent,* and *facile conception of the facts.* When once we have reached the point where we are everywhere able to detect the same few simple elements, combining in the ordinary manner, then they appear to us as things that are familiar; we are no longer surprised, there is nothing new or strange to us in the phenomena, we feel at home with them, they no longer perplex us, they are *explained.* It is a process of adaptation of thought to facts with which we are here concerned.

Economy of communication and of apprehension is of the very essence of science. Herein lies its pacificatory,* its enlightening, its refining element. Herein, too, we possess an unerring guide to the historical origin of science. In the beginning, all economy had in immediate view the satisfaction simply of bodily wants. With the artisan, and still more so with the investigator, the concisest and simplest possible knowledge that is attained with the least intellectual expenditure—naturally becomes in itself an economical aim; but though it was at first a means to an end, when the mental motives connected therewith are once developed and demand their satisfaction, all thought of its original purpose, the personal need, disappears.

(*Science of Mechanics.* E. Mach. Tr. T. J. McCormack, 1902. First pub. 1883. p. 6.)

THE SIMPLIFICATION OF PROBLEMS

Take the simplest operation considered in statics—the use of a crow-bar in raising a heavy stone, and we shall find . . .

* i.e. removing controversy.

that we neglect far more than we observe. If we suppose the bar to be quite rigid, the fulcrum and stone perfectly hard, and the points of contact real points, we might give the true relation of the forces. But in reality the bar must bend and all the extension and compression of different parts involve us in difficulties. Even if the bar be homogeneous in all its parts, there is no mathematical theory capable of determining with accuracy all that goes on: if, as is infinitely more probable, the bar is not homogeneous, the complete solution will be indefinitely more complicated, but hardly more hopeless. No sooner had we determined the change of form according to simple mechanical principles than we should discover the interference of thermodynamic principles. Compression produces heat and extension cold, and thus the conditions of the problem are modified throughout. In attempting a fourth approximation we should have to allow for the conduction of heat from one part of the bar to another. All these effects are utterly inappreciable in a practical point of view, if the bar be a good stout one; but in a theoretical point of view they entirely prevent our saying that we have solved a natural problem. . . .

(*The Principles of Science*. W. Stanley Jevons, 1874. Vol. II, p. 77.)

DIRAC ON MODERN PHYSICS AND MENTAL MODELS

The methods of progress in theoretical physics have undergone a vast change during the present century. The classical tradition has been to consider the world to be an association of observable objects (particles, fluids, fields, etc.) moving about according to definite laws of force, so that one could form a mental picture in space and time of the whole scheme. This led to a physics whose aim was to make assumptions about the mechanism and forces connecting these observable objects, to account for their behaviour in the simplest possible way. It has become increasingly evident in recent times, however, that nature works on a different plan. Her fundamental laws do not govern the world as it appears in our mental picture in any very direct way, but instead they control a substratum of which

we cannot form a mental picture without introducing irrele-
vancies. The formulation of these laws requires the use of the
mathematics of transformations. The important things in the
world appear as the invariants (or more generally the quantities with simple transformation properties) of these transformations. The things we are immediately aware of are the relations
of these nearly invariants to a certain frame of reference, usually
one chosen so as to introduce special simplifying features which
are unimportant from the point of view of general theory.

(We need not expect to understand the last two sentences
fully, but merely to note the modern physicist's awareness of
a world underlying, but not identical with, that which he
observes.)

(*The Principles of Quantum Mechanics*. P. A. M. Dirac. Oxford,
1935.)

EDDINGTON'S PARABLE OF THE FISHING-NET

Let us suppose that an ichthyologist is exploring the life of
the ocean. He casts a net into the water and brings up a fishy
assortment. Surveying his catch he proceeds in the usual
manner of a scientist to systematise what it reveals. He arrives
at two generalisations.

(1) No sea-creature is less than two inches long.

(2) All sea-creatures have gills.

These are both true of his catch, and he assumes tentatively
that they will remain true however often he repeats it.

In applying this analogy, the catch stands for the body of
knowledge which constitutes physical science, and the net for
the sensory and intellectual equipment which we use in obtaining it. The casting of the net corresponds to observation: for
knowledge which has not been or could not be obtained by
observation is not admitted into physical science.

An onlooker may object that the first generalisation is wrong.
"There are plenty of sea-creatures under two inches long, only
your net is not adapted to catch them." The ichthyologist
dismisses the objection contemptuously. "Anything uncatch-
able by my net is *ipso facto* outside the scope of ichthyological

knowledge, and is not part of the kingdom of fishes which has been defined as the theme of ichthyological knowledge. In short, what my net can't catch isn't fish." Or—to translate the analogy—"If you are not simply guessing, you are claiming a knowledge of the physical universe discovered in some other way than by the methods of physical science and admittedly unverifiable by such methods. You are a metaphysician. Bah!"

The dispute arises, as many disputes do, because the protagonists are talking about different things. The onlooker has in mind an objective kingdom of fishes. The ichthyologist is not concerned as to whether the fishes he is talking about form a subjective or objective class; the property that matters is that they are catchable.

. . . When the ichthyologist rejected the onlooker's suggestion of an objective kingdom of fishes as too metaphysical, and explained that his purpose was to discover laws (i.e. generalisations) which were true for catchable fish, I expect the onlooker went away muttering: "I bet he does not get very far with his ichthyology of catchable fish. I wonder what his theory of the reproduction of catchable fish will be like. It is all very well to dismiss baby fishes as metaphysical speculation; but they seem to me to come into the problem."

(*The Philosophy of Physical Science.* Sir Arthur Eddington. Cambridge. 1939. pp. 16–17, 62.)

CHAPTER TWENTY-THREE

The Functions of Science

What Science can do

It should be clear from Chapter VIII and the last chapter that Science is made up of observations of certain limited kinds (pp. 338-39) about such things as can be observed in that way. The observations are of an approximately known degree of certainty and they are analysed and grouped into scientific laws and theories, which together form the whole body of science. Some of the theories of science almost amount to certainties (e.g. the kinetic theory of gases) others are very conjectural (e.g. Wegener's theory of floating continents), but it is always open to us to inspect the evidence and to find out whether the answer that science gives to a question is almost certain, highly conjectural, or somewhere in between. The general public is apt to neglect this and to think that our theories of expanding universes are based on the same sure grounds as our theory of expanding railway-lines. But conjectural or certain, the scientific answer is always based on unprejudiced impersonal survey of evidence gathered in the most impartial manner, so that where information about the matters studied by science is wanted, the scientific answer is always of value. Its value is, first, that it gives a picture of an interesting, beautiful and orderly system of nature, the mere knowledge of which is a worthy goal of man's intellect: secondly, that the knowledge of the workings of material things enables us to control them.

So, first of all, science gives an intelligible picture or model of what we perceive by sense in so far as it can be expressed in terms of numbers, ratios, sizes, shapes, weights, and motions and all the qualities and units derived from these. Thus, science sums up in a rational way a great part of our relation with the external world. Some would say it sums up or could sum up all of it, but this seems to be quite unjustifiable, as is maintained

350

in the following section. So the scientific view of the world is a way of relating ourselves to it; and we enlarge ourselves by relating ourselves to the whole of the universe that man can observe and has observed by the methods of science. It also affords *material* for intellectual pleasure, wonder, joy, and the perception of unity and design in that world. These are no part of science, but the knowledge of science may evoke them in the mind of man.

Furthermore, we can deduce from this provisional working-model of the universe that we call science, the best way in which to make the external world conform to our wills. Our wills are no part of science, and science cannot tell us what we *ought* to do: it can very often, however, tell us what will be the result of doing certain things. Thus when we have settled what we wish to do, science will tell us how to do it, and what will happen when we have done it, in so far as this is within the scope of that which science includes in its study. It cannot tell us how to sculpture a beautiful statue, for beauty is not a scientific conception, but it can tell us how to make an exact reproduction of an object, because this is a process concerned with one of the quantities (distance) that science studies. As regards actual projects, science may in the future show us how to move anything anywhere at any speed. It may enable us to make an unlimited variety of new materials, of every kind. It may enable us to modify our bodies and even our nervous organisation to some extent, to cure or prevent all diseases, possibly to prolong life to an indefinite extent. It may so far increase the means of production of food, and of goods in general, that every man can have what he wants and do very little work to get it. It can increase our available stock of knowledge indefinitely, but not complete it, for there will always be another question to ask. It can increase the individual's personal knowledge greatly by presenting great tracts of experience in the form of simple laws. It can also increase the powers of tyranny over man and abuse of nature to unthinkable limits, and has already afforded man the most gigantic power to destroy his fellows. Such are the powers of science.

The Limitations of Science

Science is derived by reasoning from observations and so nothing can be found in it that is not in the observations, unless it has been illegitimately introduced in the reasoning. Thus, for example, you cannot get morals out of science. You can decide what is good or desirable and then consult science as to how to get it, but science cannot tell you what is good. Thus you may decide that 'man *should* follow the direction of his evolution', but this is just your own personal preference. Science only summarises what has been observed: the individual has to make his own decisions. Science is one kind of knowledge: knowledge is the material on which decisions are based, but knowledge does not decide.

Secondly, science summarises what has happened in the form of laws, and, assuming conditions to be unchanged, applies these to the past and future. Thus our present knowledge of dynamics and astronomy tell us that no eclipse can be total for longer than 7 min. 40 secs.: as we believe we know all the important factors of the case and that they remain nearly constant over long periods we would assert with great conviction that in the thousandth century A.D. eclipses will not exceed 7 min. 40 secs. of totality. But in the year 47,239 a dark star may come near enough to the solar system to alter the planetary orbits profoundly and so completely alter the laws governing the length of eclipses.

Past and future conditions cannot be wholly known, and so no past or future event can be certainly deduced from science. The more remote is the past and future event to be considered the less the certainty of deduction, for none of the scientific laws we adopt is known with greater certainty than the experimental error of our instruments; and the minutest change of laws with time, though now undetectable, would totally transform them in a long enough period.

Such considerations as these are important in connection with miracles. Does science tell us that water was not changed into wine at Cana of Galilee? By no means: it tells us that neither that change nor any analogous one has yet been observed, that like causes have always been found to produce

like effects, and therefore that *if* no other causes were operating
at Cana than in our laboratories, *then* the probability of the
event is very low. But this, of course, tells us nothing; for the
tenets of religion are that a Cause was operating there that does
not operate in our laboratories: this Science cannot affirm
or deny.

Those who remember that scientific laws are summaries of
observations and not legal enactments will have no difficulty
in discovering the phenomena to which they can safely be
applied.

How Science can best be applied

It will be evident that man needs a philosophy or religion to
show him what to do, and science to enable him to do it.
The former is something that science cannot provide for him,
though an accurate knowledge of the perceptible workings of
the perceptible world can provide material which can aid the
judgment in arriving at some conclusions in this matter. Thus
science indicates an orderliness and simplicity in the world,
and this must be taken into account in any religion or philosophy that purports to give an account of the origin, existence
and future of the universe. It may even be taken to indicate a
plan according to which the universe is and becomes; but of
course science does no more than present to us our summarised
perceptions of the world, and we deduce order, plan, or reason
for this, not by scientific, but by philosophical argument.

But, no doubt, whatever be the religion or philosophy we
may adopt, we shall draw the conclusion that the world as
we know it needs to be modified in order to be conformed to
that which our philosophy or religion regards as a more desirable state, and this modifying of the outer world is exactly
what science tells us how to perform. So given the will to bring
about some state of affairs—such as health, good housing, short
hours of work, general diffusion of information, abundant food
—the first necessary step is the discovery of the way in which
it can be done. To this end we have first to discover scientific
principles, then apply them in practice.

It does not follow that we should concentrate research on

to the problem we want to solve, for a great number, perhaps the majority, of first-class discoveries affecting the lives of men have been made by men who were looking for something else, but had the wits to see the importance of what they had hit on by chance. Pasteur in trying to improve beer started the chain of work that led to the discovery of the germ-theory of disease, which has saved countless lives: Röntgen in studying the conduction of electricity through gases, found X-rays: Perkin in looking for quinine, found the first aniline dye: Hertz, in confirming Clerk Maxwell's electromagnetic theory of light, discovered radio-waves. Research deliberately designed to discover some way of doing or making something is bound to be based on what we already know about that thing, and is therefore unlikely to break wholly new ground. If we require really epoch-making discoveries we must see that all departments of science are investigated without thought of the use but only of the knowledge; in the course of finding the new knowledge, new principles will be discovered. This is the task of research institutes, universities, and the like: but the application of these principles is more appropriate, perhaps, to the research laboratories of commercial undertakings, where the end to be achieved and the practical problems of production are best understood.

In order that we may bring about all the ends we desire—good or, alas, bad—we must first arrange to have as much research as possible into problems of pure science. This requires two factors, men and money.

Research workers cannot be obtained for the asking. The supply of men with the mental equipment is limited, and the best that can be done is to see that none of these are lost to less necessary professions. The Government, *i.e.* we ourselves, should see that young people of promise in science are educated at no cost to themselves and maintained during the long period of training. It should create sufficient research posts to ensure permanent occupation for them, and it should pay them a salary equivalent to that of other professional men. The result would be an increase of research, and the principles of knowledge of how to do what we want to do. The application of these principles would then be made by those with experience of practical problems—industrial chemists, engineers, medical practitioners and so on.

The Control of Science

It is obvious that Science is a source of enormous power, and that each man and nation can employ physical forces that were not dreamed of a century ago. Mental forces, if the phrase may be used, have increased no less, for the distribution of information, which is the chief spring of human action, has been made enormously more efficient and controllable. This process is not likely to stop, for it has been becoming ever more rapid for the last hundred and fifty years. The recent war (1945) shows how vast are the forces that can be put in action by men who, by their manner of using them, have shown themselves utterly unfit to be entrusted with them. Contrast the war-making and propaganda of 1945 with that of 1895. Scientific discovery has never ceased and for the last 150 years it has at least doubled its pace in each succeeding half-century. We can be confident therefore that in 1995 men's power to destroy will be many times greater than it is at present. We cannot estimate the future extension of atomic warfare. Biological discovery is likely to add its horrors, for it would seem likely that in half a century we shall have some real knowledge of epidemic plagues and the power to use them. Defence will doubtless increase, but considering solely the progress of science, it seems that if man does not restrain the desire to destroy his fellow-men, science will enable him to gratify it with ever more appalling results. Can science control its own use? If we mean by science not only the knowledge, but the workers—the answer is *Yes!* Medical men have professional ethics and they do not permit their art to be used save for the healing of man. If scientific men had the same ethics, they could save the world; for without the man of science not a wheel can turn. But will scientists try to control their science? Or will they, as to-day, consent to serve anyone who pays them? Will they always be moved by the pressure of their fellow-countrymen to use their knowledge to destroy their fellow-men in other lands? The world faces the most serious threat in history—and is making little effort to avert its doom.

Hora novissima, tempora pessima sunt; vigilemus.

Examples Concerning the Powers of Science

LAPLACE ENVISAGES SCIENCE AS A COMPLETE GUIDE TO THE FUTURE

We ought then to regard the present state of the universe as the effect of its antecedent state and as the cause of the state that is to follow. An intelligence which should be acquainted with all the forces by which nature is animated, and with the several positions at any given instant at all the parts thereof, provided that its intellect were vast enough to submit these data to analysis, would include in one and the same formula the movements of the largest bodies and those of the lightest atom. Nothing would be uncertain for it, the future as well as the past would be present to its eyes. The human mind, in the perfection it has been able to give to astronomy affords a feeble outline of such an intelligence. Its discoveries in mechanics and geometry, joined to that of universal gravitation, have brought it within reach of comprehending in the same analytical expressions the past and future states of the systems of the world. All its efforts in the search for truth cause it continually to approach the intelligence we have just conceived but from this intelligence it will ever remain infinitely distant.

(*Essai Philosophique sur les Probabilités*. Laplace. 1814.)

[Laplace realises that no human intelligence can reach his ideal. But the whole passage is based on the unconscious assumption that laws of nature that have been found to account for the facts over a period of a few centuries and within our limits of experimental error, can be expected to continue to apply with perfect accuracy for limitless centuries. This problem is a living one as appears in the next extract.]

BRIDGMAN TAKES THE OPPOSITE VIEW

"To me the most striking thing about cosmogony* is the perfectly hair-raising extrapolations† which it is necessary to make. We have to extend to times of the order of 10^{13} years and distances of the order of 10^9 light-years laws which have been checked in a range of not more than 3×10^2 years, and certainly in distances not greater than the distance which the solar system has travelled in the time, as about 4×10^{-2} light-years. It seems to me that one cannot take such extrapolations seriously unless one subscribes to a metaphysics that claims that laws of the necessarily mathematical precision *really* control the actual physical universe. For such a metaphysical claim I can find no operational meaning that would give one the faintest confidence in applying it in any concrete situation."

(P. W. Bridgman. *The Nature of Physical Theory*. Princeton University Press. 1936. p. 109.)

HERBERT SPENCER ENVISAGES A UNIVERSAL SOCIETY

Of international arbitration we must say, as of a free constitution, or a good system of jurisprudence, that its possibility is a question of time. The same causes which once rendered all government impossible have hitherto forbidden this widest extension of it. A federation of peoples—a universal society, can exist only when man's adaptation to the social state has become tolerably complete. We have already seen that in the earliest stage of civilization, when the repulsive force is strong, and the aggregate force weak, only small communities are possible; a modification of character causes these tribes, and satrapies, and *gentes*, and feudal lordships, and clans, gradually to coalesce into nations; and a still further modification

---

* Theories of the origin of the universe.

† Deductions made by applying a law to conditions outside the range of those from which it was deduced. If the population of a country in millions is 105 in 1910, 110 in 1920, 115 in 1930, and 120 in 1940, it would be an *inter*polation to deduce that it was 117·5 in 1935 but an *extra*polation to say it will be 145 in 1990 or was 95 in 1890. Obviously we assume, in making the deduction, that conditions do not change. But the more distant is the time that we consider, the more unsafe is this supposition.

will allow of a still further union. That the time for this is now drawing nigh, seems probable. We may gather as much from the favour with which such an arrangement is regarded. The recognition of its desirableness foreshadows its realization. In peace societies, in proposals for simultaneous disarmament, in international visits and addresses, and in the frequency with which friendly interventions now occur, we may see that humanity is fast growing towards such a consummation. Though hitherto impracticable, and perhaps impracticable at the present moment, a brotherhood of nations is being *made* practicable by the very efforts used to bring it about. These philanthropic enthusiasms, which the worldly-wise think so ridiculous, are essential parts of the process by which the desideratum is being wrought out. Perhaps no fact is more significant of the change going on than the spread of that non-resistance theory lately noticed. That we should find sprinkled among us, men, who from the desire to receive this ultra-humane doctrine do violence to their perceptions of what is due to themselves, cannot but afford matter for congratulation. Unsound as the idea may be, its origin is good. It is a redundant utterance of that sympathy which transforms the savage man into the social man, the brutal into the benevolent, the unjust into the just; and, taken in conjunction with other signs of the times, prophesies that a better relationship between nations is approaching. Meanwhile, in looking forward to some all-embracing federal arrangement, we must keep in mind that the stability of so complicated a political organization depends, not upon the fitness of one nation but upon the fitness of many.

(*Social Statics.* Herbert Spencer. 1851.)

The survival of humanity seems to depend on the abolition of scientific warfare, and the most hopeful means of attaining this is a measure of federation or co-operation between all nations or the merging of nations in a universal society. Since Herbert Spencer wrote this passage, nearly a century ago, war has advanced from cannon-balls and muskets to submarines, tanks and atomic bombs; while federation and co-operation has achieved only a partially successful attempt to operate a League of Nations. Can this process of co-operation be accelerated, for it seems as if destruction is winning the race? If men become either kind or sensible, so that they actively desire the good of their foreign neighbours, and no longer wish to gain a quick advantage by fraud and violence, then it might be a matter of but a few years before war was done with. If every man could be made to realise the need for this, by the same efficient means as have made him believe in the need for the supremacy of his country and the destruction of his enemies, who knows but the world might yet be saved? There is a dispassionate statement in Matt. 26, 52: *All they that take the sword shall perish with the sword*—how many more wars will be needed to teach the world this truth?

THE END

Suggestions for Further Reading

The following brief list includes only works that are obtainable, new or second-hand, at a fairly reasonable price, and which are most nearly suitable for students' reading.

GENERAL HISTORIES OF SCIENCE

A History of Science. W. C. Dampier. Cambridge University Press.
A Short History of Science. C. Singer. Oxford University Press.
A Short History of Science. F. Sherwood Taylor. Heinemann.
A Short History of Science. Sedgwick and Tyler. Macmillan.
Science Since 1500. H. T. Pledge. H.M.S.O.
A Shorter History of Science. W. C. Dampier. C.U.P.
Cambridge Readings in the History of Science. Dampier-Whetham. C.U.P.

GENERAL HISTORIES OF PARTICULAR SCIENCES

MATHEMATICS

The History of Mathematics in Europe. J. W. N. Sullivan. O.U.P.
A Short History of Mathematics. W. W. R. Ball.
History of Mathematics. D. E. Smith. Ginn, Chicago.
A Short History of Mathematics. V. Sanford. Harrap.
The History of Arithmetic. L. C. Karpinski. Rand McNally.
A Source Book in Mathematics. D. E. Smith. McGraw-Hill.

PHYSICS

A Short History of Physics. H. Buckley. Methuen.
History of Physics. Cajori. Macmillan.
Source Book in Physics. W. N. Magie. McGraw-Hill.

CHEMISTRY

A Short History of Chemistry. J. R. Partington. Macmillan.
History of Chemistry. E. von Meyer. Macmillan.
Historical Introduction to Chemistry. Lowry.
Pictorial History of Chemistry. Ferchl-Sussenguth. Heinemann.

BIOLOGY

A Short History of Biology. Singer. O.U.P.
The History of Biology. E. Nordenskiöld. Kegan Paul.
The Growth of Biology. W. A. Locy. Bell.
History of Botany. J. Reynolds Green. O.U.P.
A Short History of the Plant Sciences. Howard S. Reed. Chronica Botanica Co.
A Hundred Years of Anthropology. T. K. Penniman. Duckworth.
History of Geology and Palæontology. V. Zittel.

MEDICINE

History of Medicine. F. H. Garrison. Saunders.
History of Medicine. M. Neuburger. O.U.P.
Growth of Medicine. A. H. Buck. Yale University Press.

ASTRONOMY

A History of Astronomy. W. W. Bryant. Methuen.
A History of Physical Astronomy. R. Grant.
A Source Book in Astronomy. H. Shapley and H. E. Howarth. McGraw-Hill.

Works on Special Periods and to be Studied in Connection with Certain Chapters

CHAPTER

I. *The Legacy of Egypt.* Ed. S. R. K. Glanville. O.U.P.
II. *The Legacy of Greece.* Ed. Sir R. Livingstone. O.U.P.
   Greek Biology and Greek Medicine. C. Singer. O.U.P.
   Science in Antiquity. B. Farrington. O.U.P.
   Science and Mathematics in Classical Antiquity. J. L. Heiberg. O.U.P.
   Greek Astronomy. Sir T. L. Heath. Dent.
III. *The Legacy of Islam.* Ed. Arnold and Guillaume. O.U.P.
   Studies in the History of Mediaeval Science. C. H. Haskins.

CHAPTER

V.-VII. *Metaphysical Foundations of Modern Science.* Burtt. Kegan Paul.

Galileo and the Freedom of Thought. Sherwood Taylor. Watts.

History of Science, Technology, and Philosophy in the Sixteenth and Seventeenth Centuries. A. Wolf. Allen and Unwin.

Novum Organum. Francis Bacon. (Many editions.)

Discourse on Method. Descartes. (Many editions.)

Sceptical Chymist. Robert Boyle. Everyman Library.

VIII. *History of Science, Technology and Philosophy in· the Eighteenth Century.* A. Wolf. Allen and Unwin.

Antoine Lavoisier. D. McKie. Gollancz.

IX. *Lives of the Engineers.* Samuel Smiles. (Several vols.)

X. *English Sanitary Institutions.* John Simon. Smith Elder.

XI. *Origin of Species.* Charles Darwin.

Life of Erasmus Darwin. E. Krause.

XII. *Modern Cosmologies.* H. Macpherson. O.U.P.

General Astronomy. Spencer Jones. Arnold.

XIII. *Michael Faraday, His Life and Work.* S. P. Thompson.

XIV. *Development of Transport in Modern England.* W. T. Jackman.

XVII. *The Principles of Science.* S. Jevons.

The Grammar of Science. Karl Pearson. Black.

Philosophy of Physical Science. Eddington. O.U.P.

XVIII. *The Fourfold Vision.* Sherwood Taylor. Chapman and Hall.

INDEX

THE NORTON LIBRARY

Gorer, Geoffrey, and John Rickman, M.D. *The People of Great Russia:* A Psychological Study. N112

Gosse, Edmund. *Father and Son.* N195

Graves, Robert and Alan Hodge. *The Long Week-end:* A Social History of Great Britain, 1918-1939. N217

Hamilton, Edith. *Spokesmen for God.* N169

Hamilton, Edith, Tr. and Ed. *Three Greek Plays.* N203

Hamilton, Edith. *Witness to the Truth:* Christ and His Interpreters. N113

Harrod, Roy. *The Dollar.* N191

Hawthorne, Nathaniel. *The Blithedale Romance.* Introduction by Arlin Turner. N164

Hinsie, Leland. *The Person in the Body.* N172

Homer. *The Iliad, A Shortened Version.* Translated and Edited by I. A. Richards. N101

Horney, Karen. *Are You Considering Psychoanalysis?* N131

Huxley, Aldous. *Texts and Pretexts.* N114

James, William. *Talks to Teachers.* Introduction by Paul Woodring. N165

Kelly, George A. *A Theory of Personality:* The Psychology of Personal Constructs. N152

Keynes, John Maynard. *Essays in Biography.* N189

Keynes, John Maynard. *Essays in Persuasion.* N190

Knight, G. Wilson. *The Christian Renaissance.* N197

Lang, Paul Henry, Editor. *Problems of Modern Music.* N115

Lang, Paul Henry, Editor. *Stravinsky: A New Appraisal of His Music.* N199

Lawrence, T. E. *The Mint.* N196

Leavis, F. R. *Revaluation:* Tradition and Development in English Poetry. N213

Lunt, Dudley C. *The Road to the Law.* N183

Mackenzie, Henry. *The Man of Feeling.* Introduction by Kenneth C. Slagle. N14

Mackinder, Halford J. *Democratic Ideals and Reality.* New introduction by Anthony J. Pearce. N184

Moore, Douglas. *A Guide to Musical Styles:* From Madrigal to Modern Music. N200

Moore, Douglas. *Listening to Music.* N130

Moore, George. *Esther Waters.* Intro. by Malcolm Brown. N6

Morey, C. R. *Christian Art.* With 49 illustrations. N103

Morrison, Hugh. *Louis Sullivan:* Prophet of Modern Architecture. Illustrated. N116

Nicolson, Marjorie Hope. *Mountain Gloom and Mountain Glory:* The Development of the Aesthetics of the Infinite. N204

Ortega y Gasset, José. *Concord and Liberty.* N124